Praise for Lawrenc

ATOM

A SINGLE OXYGEN ATOM'S ODYSSEY
FROM THE BIG BANG
TO LIFE ON EARTH . . . AND BEYOND

"The startling truth is that the trillions and trillions of elemental particles that are the essence of us — the atoms of hydrogen, carbon, oxygen, and so on that are our human underpinnings, in every fingernail and eyelash, every muscle and blood cell and bone — are restless galactic travelers that belong to us only temporarily. . . . *Atom* weaves cutting-edge theories of cosmology with nuggets from evolutionary biology, paleontology, and genetics." — John Mangels, *Cleveland Plain Dealer*

"Through the life's arc of a single oxygen atom, the story of matter is told. *Atom* encompasses the twentieth-century revolution in physics, the men and machines that delve into the mysteries of the universe, as well as the epic journey of this atomic voyage. The story of this atom becomes our story, its many lives our lives." — *Astronomical Society of the Pacific*

"Surpassing even Blake's vision of the world in a grain of sand, Krauss offers readers the entire cosmos in a mere atom." — Bryce Christensen, *Booklist* (starred review)

"Through clear exposition, Krauss imparts a solid understanding of astrophysics." — *Science News*

"Meticulously detailed. . . . Through an exploratory discussion of how life may have unfolded, the author's ripe imaginative powers and literary prowess come into play. Krauss presents a wealth of information that covers a range of disciplines (such as geophysics, biology, and paleontology) and concludes with a glimpse of the future, where the forces that spawned life will destroy it." — *Publishers Weekly*

ATOM

ATOM

A SINGLE OXYGEN ATOM'S ODYSSEY
FROM THE BIG BANG
TO LIFE ON EARTH . . .
AND BEYOND

LAWRENCE M. KRAUSS

BACK
BAY
BOOKS

LITTLE, BROWN AND COMPANY · BOSTON NEW YORK LONDON

Originally published in hardcover by Little, Brown, April 2001
First Back Bay paperback edition, May 2002

The author is grateful for permission to quote from the following sources: Excerpt from "East Coker" in Four Quartets, copyright 1940 by T. S. Eliot and renewed 1968 by Esme Valerie Eliot, reprinted by permission of Harcourt, Inc.; "Brotherhood (Hermandad)" by Octavio Paz, translated by Eliot Weinberger, from Collected Poems 1957–1987, copyright © 1986 by Octavio Paz and Eliot Weinberger. Reprinted by permission of New Directions Publishing Corp.

Library of Congress Cataloging-in-Publication Data
Krauss, Lawrence Maxwell.
Atom : a single oxygen atom's odyssey from the Big Bang to life on Earth . . . and beyond / by Lawrence M. Krauss — 1st ed.
p.   cm.
Includes index.
ISBN 0-316-49946-3 (hc) / 0-316-18309-1 (pb)
1. Cosmology.   I. Title.
QB981 .K77 2001
523.1 — dc21   00-055795

10 9 8 7 6 5 4 3 2 1

Designed by Iris Weinstein

Q-MART

Printed in the United States of America

TO KATE, FOR 21 YEARS OF PATIENCE

I am a man: little do I last
and the night is enormous.
But I look up:
the stars write.
Unknowing I understand:
I too am written,
and at this very moment
someone spells me out.

OCTAVIO PAZ

CONTENTS

CONTENTS

ATOM

THE CITY
ON THE EDGE
OF FOREVER

As you wander past the Hôtel de Matignon, the official residence of the French prime minister, past the art and antiques shops that crowd together amid the bustle and color of the 7th arrondissemont in Paris, you arrive at the courtyard of a grand eighteenth-century estate, whose walls protect a garden enclave from the traffic, noise, and concerns of the outside world. Throughout the gardens and the mansion located at their center, you can gaze upon the works of one of the nineteenth century's greatest sculptors, Auguste Rodin.

If you pay the entry fee and walk inside the villa, up the grand staircase to the upstairs foyer overlooking the immaculate gardens below, you will come face to face with the miracle by which solid rock is transformed into the sensuous outlines of the human form. While sculptors throughout the ages have created beautiful images in media ranging from rock and bronze to wood and glass, the uniqueness and majesty of Rodin's art lie in part in the striking juxtaposition of form and formlessness. It seems as if the rock itself is giving birth to the tender and sometimes tragic human shapes that rise up from its foundation: a couple locked in a caress, a nymph at rest, humanity cradled in a huge guardian hand. Whenever my eyes move from the rough-hewn edges to the smooth shapes within, my mind also begins to wander, but over a far broader horizon. I cannot help picturing this transformation as an allegory for our own long journey out of nothingness. As I touch the cool and solid marble, and marvel at a couple locked in an ecstatic and apparently eternal embrace, I ask myself whether this too is an illusion, whether anything is eternal, and where our own future lies. And I think about my own remarkable fortune to be a living, cognizant creature who can make such speculations at a time when some of nature's most perplexing mysteries may be yielding to our persistent push. Such is, I suppose, the power of great art — to transport you beyond your immediate concerns and free your mind and spirit to wander.

The world's religions too speak of creation and transformation, of life and death and sometimes resurrection. The cycle of life — birth, death, and birth again — has occurred with clocklike regularity, on scales ranging from minutes to millennia, over the course of eons on Earth. But together, all these many lives and deaths represent merely a snapshot in cosmic time. The universe we understand existed for almost twice as long before Earth was formed as it has existed since the cosmic bits of rock and dust first coalesced together around a medium-size star at the edge of the Milky Way galaxy. And we know for certain that the universe will continue to exist, largely unchanged, for at least twice as long again, long after our own sun has puffed up and swallowed the Earth, and before it in turn slowly dies, like an ember in a fireplace losing its glow in the dark at the end of a long winter's night.

We are said to go from ashes to ashes, dust to dust. But though our nature compels us to think of our own experience as the defining feature of existence, it is not. All the while, the fundamental protagonists in the drama of life are the very atoms that make up our bodies. They may experience what we all desire: a chance at immortality.

This book tells their story.

Like all good drama, this story is not about *all* atoms, because atoms, like people and dogs, and even cockroaches, have individual histories. Rather, this is a story about one atom in particular, an atom of oxygen, located in a drop of water, on a planet whose surface is largely covered by water but whose evolution is for the moment dominated by intelligent beings who live on land. It could, at the present moment, be located in a glass of water you drink as you read this book. It could have been in a drop of sweat dripping from Michael Jordan's nose as he leapt for a basketball in the final game of his career, or in a large wave that is about to strike land after traveling 4,000 miles through the Pacific Ocean.

No matter. Our story begins before the water itself existed, and ends well after the planet on which the water is found is no more, the myriad human tragedies of the eons perhaps long forgotten. It is a story rich in drama, and poetry, with moments of fortune and remarkable serendipity, and more than a few of tragedy.

As I embark on this story, I cannot help reflecting on how many times my mother admonished me as a child, "Don't touch that, you don't know where it's been!" She would have been surprised. . . .

PART ONE DIVINE WIND

The world becomes stranger, the
    pattern more complicated
Of dead and living. Not the intense
    moment
Isolated, with no before and after,
But a lifetime burning in every moment

T. S. ELIOT

1.
THE UNIVERSE
IN AN ATOM

Many are called, but few are chosen.
MATTHEW 22:14

In the year 1281, the second Mongol invasion of Japan began, and ended. The invaders were defeated as much by the force of nature as by the Japanese warriors, as the Mongol ships suffered grievous losses due to the *Kamikaze*, or "divine wind." This routed the invaders and boosted Japanese pride in their island's invincibility, much as the storms that helped repel the Spanish Armada from British shores 307 years later — immortalized in a commemorative medal with the words "God Blew, and they were scattered" — helped affirm the sense of Divine Right harbored by Mother England for centuries thereafter.

Those Mongol ships that survived the crossing of the Sea of Japan may have noticed the range of mountains that rise sharply from the water near the town of Toyama. These are known by some as the Japanese Alps — a popular skiing attraction today. Deep below these snowy peaks, where the sun never shines, indeed has never shone, may lie the secret of our existence, forged from a fiery wind, not necessarily divine, but more intense than any that has ever swept the Earth and as old as creation itself.

In the deep Mozumi mine in the town of Kamioka lies an immense tank of pure, clear water, recycled daily to remove contaminants. Forty meters in diameter and over 40 meters high, the Super-Kamiokande detector, as it is known, contains 50,000 tons of water — enough to quench the thirst of everyone in a city the size of Chicago for a day. Yet this device, located in a working mine, is maintained with the spotless cleanliness of an ultra-purified laboratory clean room. It has to be. The slightest radioactive contaminants could mask the frustratingly small signal being searched for by the scores of scientists who monitor the tank with 11,200 phototubes — eerily resembling television tubes — lining the outside of the tank. If the scientists' attention wavers for even a second, they could miss an event that might not occur again in the lifetime of the detector, or the scientists. A single event could explain why we live in a universe of matter, and how long the universe as we know it may survive. The signal they are searching for has been hidden for at least 10 billion years — older than the Earth, older than the sun, and older than the galaxy. Yet compared to the timescale of the process behind the event being searched for, even this stretch is just the blink of a cosmic eye.

We are about to embark on a journey through space and time, traversing scales unimaginable even a generation ago. A tank of water located in the dark may seem an odd place to begin, but it is singularly appropriate on several grounds. The mammoth detector contains more atoms — by a factor of 100 billion or so — than there are stars in the visible universe. Yet amid the 10^{34} (1 followed by 34 zeros) or so atoms in the tank is a single oxygen atom whose history is about to become of unique interest to us. We do not know which one. Nothing about its external appearance can give us any clue to the processes that may be occurring deep inside. Thus we must be ready to treat each atom in the tank as an individual.

The vast expanse of scale separating the huge Super-Kamiokande tank and the minute objects within it is a prelude to a voyage inward where we will leave all that is familiar. The possible sudden death of a single atom within the tank might hearken back to events at the beginning of time.

But beginnings and endings are often inextricably tied. Indeed, each Sunday one can hear proclaimed loudly in churches across the land: "As it was in the beginning, is now, and ever shall be, world without end." But do those who recite these words expect that they refer to our world of human experience? Surely not. Our Earth had a beginning. Life had a beginning. And as sure as the sun shines, our world will end.

Can we nevertheless accept this prayer as metaphor? Our world will end, but our world is merely one of a seemingly infinite number of worlds, surrounding an unfathomable number of stars located in each of an even larger number of galaxies. This state of affairs was suspected as early as 1584 when the Italian philosopher Giordano Bruno penned his *De l'infinito universo e mondi.* He wrote:

> There are countless suns and countless earths all rotating around their suns in exactly the same way as the seven planets of our system. We see only the suns because they are the largest bodies and are luminous, but their planets remain invisible to us because they are smaller and non-luminous. The countless worlds in the universe are no worse, and no less inhabited than our Earth.

If, in the context of this grander set of possibilities, we contemplate eternity, what exactly is it that we hope will go on forever? Do we mean life? Matter? Light? Consciousness? Are even our very atoms eternally perdurable?

And so that is ultimately why our journey begins in the water in this dark mineshaft. If we explore deeply enough into even a drop of water, perhaps located in the Super-Kamiokande tank, we may eventually make out the shadows of creation, and the foreshadows of our future.

The water is calm, clear, and colorless, but this apparent serenity is a sham. Probe deeper — plop a speck of dust into a drop of water under a microscope, say — and the violent agitation of nature on small scales becomes apparent. The dust speck will jump around mysteriously, as if alive. This phenomenon is called Brownian motion, after the Scottish botanist Robert Brown, who observed this motion in tiny pollen grains suspended in water under a microscope in 1827, and who at first thought that this exotic activity might signal the existence of some hidden life force on this scale. He soon realized that the random motions occurred for all small objects, inorganic as well as organic, and he thus discarded the notion that the phenomenon had anything to do with life at all. By the 1860s, physicists were beginning to suggest that these movements were due to internal motions of the fluid itself. In his miracle year of activity, 1905, Albert Einstein proved, within months of his famous paper on relativity, that Brownian motion could be understood in terms of the motion of the individual bound groups of atoms making up molecules of water. Moreover, he showed that simple observations of Brownian motion allowed a direct determination of the number of molecules in a drop of water. For the first time, the reality of the previously hidden atomic world was beginning to make itself manifest.

It is difficult today to fully appreciate how recent is the notion that atoms are real physical entities, and not mere mathematical or philosophical constructs. Even in 1906, scientists did not yet generally accept the view that atoms were real. In that year the renowned Austrian physicist Ludwig Boltzmann took his own life, in despair over his self-perceived failure to convince his colleagues that the world of our experience could be determined by the random behavior of these "mathematical inventions."

But atoms *are* real, and even at room temperature they live a more turbulent existence than a farmhouse in a tornado, continually pulled and pushed, moving at speeds of hundreds of kilometers an hour. At this rate a single atom could in principle travel in 1 second a distance 10 trillion times its own size. But real atoms in materials change their direction at least 100 billion times each second due to collisions with their neighbors. Thus in the course of one minute, a single water molecule, containing two hydrogen and one oxygen atoms, might wander only one-thousandth of a meter from where it began, just as a drunk emerging from a bar might wander randomly back and forth all night without reaching the end of the block on which the bar is located.

Imagine, then, the chained energy! A natural speed of 100 meters per second is reduced to an effective speed of one-thousandth of a meter per minute! The immensity of the forces that ensure the stability of the world of our experience is something we rarely get to witness directly. In fact, it is usually reserved for occasions of great disaster.

You can get some feeling for the impact that tiny atoms can have on one another by inflating a balloon and tying the end, then squeezing the balloon between your hands. Feel the pressure. What is holding your hands back, stopping them from touching? Most of the space inside the balloon is empty, after all. The average distance between atoms in a gas at room temperature and room pressure is more than ten times their individual size. As the nineteenth-century Scottish physicist James Clerk Maxwell, the greatest theoretical physicist of that time, first explained, the pressure you feel is the result of the continual bombardment of billions and billions of individual molecules in the air on the walls of the balloon. As the molecules bounce off the wall, they impart an impulse to the wall, impeding its natural tendency to contract. So when you feel the pressure, you are "feeling" the combined force of the random collisions of countless atoms against the walls of the balloon.

Although this *collective* behavior of atoms is familiar, the world of our direct experience almost never involves the behavior of a *single* atom. But attempting to visualize the world from an atomic perspective opens up remarkable vistas, and gives us an opportunity to understand more deeply our own circumstances. The eighteenth-century British essayist Jonathan Swift recognized the inherent myopia governing our worldview when he penned *Gulliver's Travels*, which noted that the rituals and traditions of any society may seem perfectly rational for one who has grown up with them. Swift's Lilliputians fought wars over the requirement that eggs be broken from their smaller ends. From our vantage point, the requirement seems ridiculous. The same may be true for our view of the physical world, which is colored by a lifetime of sensory experience.

And so, as we approach the beginning of our oxygen atom's journey forward, we have to stretch our minds in the tradition of Swift. The atoms getting thrashed today in a drop of water may have a hard life, but this can't even begin to compare to the difficulties associated with their birth. To imagine these moments, we must go back to a time before water existed in the universe. We must venture back to when things were vastly more violent, back to a time more than 10 billion years ago, and perhaps less than 1 billionth of a billionth of a second after the beginning of time itself. We must visualize the universe on a scale that is so small, words cannot capture it. Indeed, we must go back to a time when there *were* no atoms . . . or Eves.

We begin when what is now the *entire* visible universe of over 400 billion galaxies, each containing over 400 billion stars, each 1 million times more massive than the Earth, encompassed a volume about the size of a baseball.

The simplicity of this statement belies its outrageousness. It is impossible to intuitively appreciate this era by making the leap from here to there in one giant step. But it *is* possible to imagine a series of smaller steps, each of which itself pushes the limits of visualization, but each of which gets us closer to fathoming the truly extreme environments we are about to enter.

Our first step begins with our own sun. Almost a million times as massive as the Earth, at its center the temperature is almost 15 million degrees, cooling by more than a factor of 1,000 at the surface to a mere 6,000 degrees, about twice the temperature of boiling iron. Nevertheless, the sun's average density is only marginally greater than that of water, not much different than the average density of the Earth, in fact. If we squeeze the sun in radius by a factor of 10, so that it is now 10 times the radius of the Earth, it is now much denser than any planet in the solar system. A teaspoon of its material would, on average, now weigh several pounds. Compress the sun by an additional factor of 10. Now the size of the Earth, with a mass 1 million times as great, each teaspoon of its material weighs several tons. Compress the sun now by another factor of 1,000. It is now about 6 kilometers in radius, the size of a small city. A single teaspoon of its material weighs 1 billion tons! (The amount of work required to perform this feat of compression, by the way, is equivalent to the total radiant energy released by the sun over the course of 3 billion years!)

At this density, the atoms in the sun lose their individual identity. Under normal conditions, a single atom is composed of a dense nucleus, made up of the elementary particles called protons and neutrons, which are themselves made up of smaller fundamental particles called quarks. The nucleus contains more than 99.9 percent of the total mass of the atom. It is surrounded by a "cloud" of electrons that occupy a space more than 10,000 times larger in radius than the nucleus but carrying almost none of the mass of the atom.

By "cloud" I actually mean nothing of the sort. "Cloud" is simply a name we give to the electron distribution because we have no really appropriate label. It is impossible to describe in words what the electrons "do" as they surround the nucleus. At this scale they are described by the laws of quantum mechanics, under which material objects behave completely unlike they do on human scales so that our normal experience is no guide whatsoever. Individual elementary particles such as electrons do not behave like "particles." They are not localized in space when they are orbiting the nucleus, as planets are when they orbit the sun — rather, they are "spread out." I say this even though we know that electrons can, under certain carefully controlled conditions, be localized on scales so small that we have not yet been able to put a lower limit on their intrinsic size, with no evidence whatsoever of any internal structure. Our language, derived from our intuitive experience of the world, has no place for such behavior.

But the electrons in an atom are not spread out over all space, merely in a volume approximately 1,000 billion times larger than the volume of the nucleus. When we compress the sun to the size of downtown Washington, D.C., we squish the atoms to the point where their electron clouds are essentially pushed *inside* the nuclei, which in turn are touching each other. The entire mass of the sun is then essentially like one huge atomic nucleus.

(As bizarrely unrealistic as such a scenario for an object like the sun may seem, it actually happens about a hundred times every second in the visible universe. In our own galaxy, about once every thirty years the inner core of a star ends its life in such a state after a massive stellar explosion — a supernova — of the type that created us.)

Let us keep on compressing. Take this gigantic solar atomic nucleus of mass 10^{56} times the mass of a hydrogen nucleus, and compress it further by another factor of 100,000, so that a single teaspoonful of material now weighs a million billion billion tons — the mass of 1,000 Earths! The sun is now the size of a basketball.

However, there are about 400 billion suns in our galaxy, and at least as many galaxies in the visible universe. Even if every star was compressed down to the size described above and all the stars in all the galaxies were packed closely together, they would still encompass a volume as large as that of the Earth. (Implying, by the way, in case it ever proves useful to you to know it, that one can fit as many basketballs inside the Earth as there are stars in the visible universe.)

We have one more large step to take. Compress all of this mass, 160,000 billion billion times the mass of the sun, down by another factor of 10 million in radius. The matter in the entire presently visible universe is now contained in a space the size of a baseball. The *mass* of a teaspoonful of this matter alone equals as much as a million galaxies, containing a total mass of a billion billion times that of our sun! In the space traditionally occupied by a *single atomic nucleus,* the amount of matter contained would be more than enough to construct all of New York City! In the space traditionally occupied by a single atom, including the region in which the electrons normally orbit, the amount of matter would be almost the mass of the entire Earth!

These numbers may seem staggering, but they do not tell the whole story. In fact, they miss the most important part of it. As one compresses matter, the energy exerted heats the material up. A larger and larger fraction of the total energy of a closed system is contained in the radiant energy emitted and absorbed by the hot particles. Well before the whole system is compressed to the unfathomable levels I have described above — in fact, when the observable universe is compressed by merely a factor of 10,000, about a million light-years across — its energy would be dominated not by matter, but by the energy of *radiation.*

The radiation at this point is so hot and dense that it beats out the gravitational pull of all 160,000 billion billion stars! But by the time we compress the visible universe down to the size of a baseball, the fraction of the total energy associated with the mass of all the matter making up all galaxies today is only about 10^{-25}, or about 1 part in 10 million billion billion! (This radiation has a huge pressure and it does work on an expanding universe, so that after a few thousand years, its energy dwindles away and becomes negligible, leaving just the matter contribution to dominate the universe today.) Thus, while in the region normally occupied today by a single atom the matter contained at that time would have a rest mass comparable to that of the Earth, the actual amount of energy contained in this region, including radiation energy, would have been much larger. In fact, it would correspond to the energy of *the entire presently visible universe!*

The universe in an atom!

Let's pause and reflect on our voyage. Even after the baby steps, it is still mind-boggling to try to picture what conditions are like when each atomic volume contains an amount of energy equivalent to that contained in our whole visible universe today. But you may wonder whether it is even worth trying. After all, under such conditions the whole meaning of "atoms," the protagonists of our story, dissolves. How can we connect individual entities like the oxygen atoms that help make up the molecules of our DNA with anything in that incredible morass?

You also might have wondered why, if we are going to go back this far, we don't go back all the way, and begin our story at the infinitely dense Big Bang itself. Let's address this second concern first. The reason we do not take our story all the way back to t=o is that this instant is still shrouded in mysteries beyond our scientific purview, so there is nothing concrete to say. But we do not think we have to go all the way back to t=o in order to understand the origin of our atoms. We believe that the Super-Kamiokande experiment, or a larger one that may follow it, may allow us to infer the events that would have had to occur at the precise moment when the existence of atoms in our universe first became a real possibility. And, to respond to the first concern, that moment occurred very early in the history of the universe. It is appropriate to argue that each atom in our bodies began life precisely then, even though atoms themselves would not exist for what would seem like an eternity at that time.

Although no events have yet been observed in the Super-Kamiokande tank that would let us re-create with some certainty the events at that time, we know that a specific, if subtle, series of events had to occur in that primordial baseball in order for our oxygen atom to exist today. So subtle and rare, in fact, that had anyone been around then to notice what was taking place, they probably wouldn't have.

Indeed, it seems that without an early series of rare events — at least as rare as a single person buying two winning lottery tickets for two different state lotteries in the same year — no one should be around today to celebrate creation, or lotteries.

Nevertheless, there is a maxim I am constantly reminded of in my work: Because the universe is big and old, no matter how unlikely something is, if it can happen it will happen. Accidents more remote than anything that might occur during our lifetime occur every second *somewhere* in the vast reaches of the cosmos. The most important question of modern science, and perhaps theology as well, is then: Are we merely one such accident?

Because Super-Kamiokande has not yet given us the empirical evidence we need to infer precisely what series of events occurred at this early time, we only know that some specific challenges, which I shall describe, had to have been met in order for our oxygen atom to exist today. In this sense the story of our atom takes on a *Rashomon*-like quality. In his famous film, Akira Kurosawa followed three different versions of the same event, a rape and murder, as remembered by three participants. Because of their different vantage points, and their different past experiences, each describes a different story. None is universally accurate, but each contains at least a germ of truth.

If atoms could speak, each would have a different story to tell. But we expect that the beginning of all these stories, when our universe could fit inside an atom, would be the same. Its rough outline has begun to emerge over the past century as scientists have carefully recorded and analyzed the signals nature has provided. The event we await at Super-Kamiokande or an experiment that may follow it will, we hope, nail the details. Until then, the following story is guaranteed to contain at least a germ of truth.

2.
THE RIGHT STUFF

*In order for the wheel to turn, for life to
be lived, impurities are needed . . .*

PRIMO LEVI

A simple accident often determines the difference between life and death. This can occur in art, as in the
1999 film *Sliding Doors*, where the fact of missing a subway train
changes the course of a young woman's future, or in real life, as when
a friend of mine missed TWA Flight 800 to Paris on July 17, 1996, in
order to visit his father in the hospital, and thus avoided incineration
in the sky.

The events near the beginning of time that immediately preceded the
birth of our atom may seem innocuous enough. But a slight alteration in
the initial conditions, and atoms, and the cosmos as we know it, would
not exist today. Just as fictional heroes, from Shakespeare's Hamlet to
Heller's Yossarian, are subject to compelling historical forces, often beyond their control, our inanimate hero is dependent upon the accidents
and vicissitudes of cosmic history.

When I speak of "accidents" it may sound as if I have given up any pre-
tense of scientific accuracy. In fact, however, predictable accidents are
the basis of essentially all modern scientific inquiry. In our laboratories
today, we literally wait for accidents — except we stack the deck, creating favorable conditions according to which the laws of probability must

play themselves out, and we wait and watch. Sometimes this involves building mammoth tanks of water in deep mines. Sometimes it involves building single machines larger and more complex than anything before created by humanity. In such machines we re-create, for a brief instant, certain features of the early moments of the Big Bang.

Located between the Jura Mountains to the north and Lake Geneva and the Alps to the south is the Geneva International Airport. If you are lucky and the low cloudbank which sometimes hides the valley is absent, then just before landing there you may glimpse a cluster of buildings less than a mile northwest of the airport. Like the tip of an iceberg, the central administration buildings of the European Laboratory for Particle Physics, or CERN (the French acronym for the original, now politically incorrect name, European Council for Nuclear Research), belie a far more impressive structure hidden below the surface. The picturesque farmland and small hamlets dotting the French–Swiss border do not betray any evidence of one of the largest tunnels in the world. Traveling 26 kilometers in a vast circle ranging from about 50 to over 100 meters below ground is the CERN large electron–positron (LEP) collider ring, soon to house a new machine, the large hadron collider (LHC).

To visit the CERN laboratory, or any major particle accelerator facility today, is to feel like Gulliver entering Brobdingnag, the land of the giants. Every object seems out of scale with mere human dimensions. The LEP tunnel itself, for example, while 26 kilometers long, is wide enough inside to easily accommodate a modern sport utility vehicle or two, should it ever be turned into an underground racetrack. The tunnel is accessible from the surface at one of four different laboratory locations. At each site, one of four mammoth particle detectors can be found. Each is the size of an apartment building built deep below ground, contained in experimental halls that could dwarf the stage at Radio City. Each behemoth is constructed from thousands of separate components fabricated by hundreds of physicists and technicians hailing from dozens of countries scattered across the globe. And each is built to a precision of fractions of a millimeter. The M.I.T. physicist Vicki Weisskopf dubbed such devices "the Gothic cathedrals of the twentieth century."

From the start of construction to the completion of the first experiment, a decade can easily pass. This is in sharp contrast to the timescale of the processes being investigated, which occur in less than a billionth of a billionth of a second. Such experiments repeatedly re-create and measure, for a fraction of a fraction of a second, the conditions and interactions of matter and energy, including, as we shall see, the extremely rare events which may have been last experienced in our universe 10 billion years ago.

The LEP collider imparted energies to elementary particles that were far beyond those produced anywhere else in our galaxy, except perhaps in the shock wave from an exploding star, or in the final collapse of a gigantic black hole. Charged particles were accelerated by electric and magnetic fields around the tunnel so that they traverse the Swiss–French border (without passports!) near the airport and then again at the base of the Jura Mountains 10,000 times each second. In the process, they achieved energies of almost 1 million times the energy they store at rest.

Yet these gargantuan values are still 1 million million times smaller than the average energy carried by *every* particle when the universe was the size of a baseball. The collisions between individual particles at that time were so energetic that to re-create them with present technology one would have to build particle accelerators with a circumference bigger than the circumference of the moon's orbit around the Earth!

The voyage from LEP to the primordial universe is more than a voyage back in time. It is a voyage in scale that helps take us and our atoms much of the way from Brobdingnag to Lilliput. As out-of-this-world as the colossal detectors in the LEP experimental halls may seem, their mismatch to normal human scales is inconsequential compared to the degree to which the scales of activity in our current universe dwarf those that were relevant when atoms were conceived.

The huge densities and temperatures at that time are reflected in an equally matched subatomic ferocity. Let us return to the barrage of atoms in a drop of water again, as seen under the microscope. The Brownian jumps of a dust speck are produced by the collective collisions of billions of jiggling atoms in the water, each momentarily traveling at hundreds of kilometers an hour. But *a single subatomic elementary particle* in the primordial gas when the universe was the size of a baseball carried enough energy to, in a single collision, knock the same dust speck not only right out of the water, but out of the Earth! If we tried to ratchet up such a collision to a human scale, it would be like accelerating a rocket in space to such a high speed that upon colliding with the moon, it could kick it right out of the solar system. (Of course, the moon would first actually break apart — as would the dust speck in the subatomic analogue — but that is not the point.) Gulliver never witnessed phenomena as foreign as this in all his travels!

Not only did such incredibly energetic collisions occur in the early universe, they occurred often and everywhere. Remember that in a region the size of a single present-day atomic nucleus, there were then more than enough particles to comprise 1 billion billion billion billion nuclei. Moreover, the collision rate was so high that in *1 second* each particle in the early proto-universe would have been able to engage in more collisions than there are grains of sand on Earth.

But of course 1 second is an eternity compared to the actual age of the universe at that time. Indeed, 1 second is far more than 1 trillion times longer, when compared to the age of the universe at that time, than the age of the universe is today, when compared to 1 second.

So these are the conditions when the gist of our oxygen atom came to be, when *nothing* became *something*.

How can I say "nothing" when there in fact was orders of magnitude more energy in a volume the size of a head of a pin at that time than is contained in our entire observable universe today? The point is there *was* a lot of stuff, but not the *right* stuff.

In spite of the mismatch between the mini-world created momentarily during the collisions at CERN and the phenomenal collisions in the very early universe, they have one thing in common. In both, energy is directly converted into mass, and vice versa — a striking example of the verity of Einstein's theory of relativity. When the universe was the size of a baseball, the energies involved in the collisions of pairs of particles were so great that 1 billion billion newly created particles could, in principle, spew out of the collision of just 2 colliding energetic electrons. And collisions were occurring so fast that no single particle preserved its identity for long: electrons smashed together to make quarks, and quarks smashed together to make particles of radiation, photons, and all of these smashed together to make unknown particles that may last have existed in nature when the universe was less than a billionth of a second old. How can we make sense of such a mishmash?

This is precisely the problem faced by elementary-particle physicists as they attempt to explore the fundamental laws of physics at accelerators. When we bang together two beams of particles in the LEP collider, or in its higher-energy cousin at Fermilab, near Chicago, we create a host of new particles in each collision, particles created out of pure energy. If not billions, then at least hundreds of new elementary particles are created as collision products. There may be a million collisions per second when the beams at Fermilab interact, each producing hundreds of particles. To simply record these events on disk would have exhausted the greatest supercomputers even a decade ago. In fact, one of the reasons that modern high-powered computer clusters were first tested at accelerator laboratories is that these were the first places they were needed.

It turns out that we can make sense of the resulting mess not by attempting to record every feature of the flotsam and jetsam, but by homing in on certain features deemed to be important. In such a way, for example, we find that for every million incident protons that are smashed into a target, amid the resulting menagerie of particles we might find a single antiproton — the nucleus of the lightest atom of antimatter, antihydrogen.

In a universe made of matter, antimatter appears to be the ultimate villain. Antimatter doesn't naturally exist on Earth in abundance for the simple reason that if it did, we wouldn't be around today to know about it. When a particle of antimatter encounters its corresponding particle of matter, the two can annihilate completely, leaving only pure energy in their wake. A single kilogram of antimatter coming in contact with a kilogram of corresponding matter could produce an explosion more powerful than any explosion humans have ever created.

The very word *antimatter* conjures up visions of exotic science fiction fantasies. But antimatter really isn't so strange. The chief distinction between particles and antiparticles is that, like a European versus a Lilliputian, we are used to seeing one and not the other.

It may seem facetious to suggest that antimatter is no less normal than matter is, but from a fundamental perspective, this is the case. Antimatter and matter are inextricably tied together like day and night. The possibility of existence for one requires the possibility of existence of the other. The theory of relativity and that other pedestal of twentieth-century physics, quantum mechanics, together imply that every type of elementary particle in nature must have a kind of alter ego with precisely the same mass, but with opposite electric charge. The antiparticle of an electron, a positron, has positive charge, and the antiparticle of the positively charged proton, an antiproton, is negatively charged.

When this prediction implying matter–antimatter duality first arose from an equation that the British physicist Paul Dirac wrote down in 1931 in his attempt to tie together relativity and quantum mechanics, no one took it seriously, least of all the developer of the equation. Amazingly, within two years of the prediction that antimatter should exist, a positron was observed amid the debris produced by one of the billions and billions of cosmic-ray particles bombarding the Earth from space every second. Dirac was said to have uttered "My equation was smarter than I was!"

Even what we call matter and antimatter is arbitrary, just as what we choose to call positive and negative electric charge is a matter of human convention. Two hundred years ago Benjamin Franklin decided to label a certain quantity "positive charge," although it later turned out that the principal carrier of electric current, the electron, has the opposite and therefore negative charge. But once we make the decision about what to call positive and what to call negative, we have to stick to it so that our physical descriptions remain unambiguous. If we had it to do all over again, it would make sense for us to call electrons positively charged, so negative signs wouldn't keep cropping up when we discuss the flow of electric current.

Now we can get to the key question: If what we call matter and antimatter is arbitrary, why do we appear to live in a universe made of one, and not the other? Put another way, if the *universe* had it all to do over again, would it be made of matter, or antimatter, or both, the next time around? If stars were made of antiprotons and positrons, instead of protons and electrons, these would join together to form antihydrogen atoms, which could then fuse together under high temperatures and pressures to create antihelium atoms. Moreover, antihydrogen would emit exactly the same set of colors of visible light as hydrogen does when you heat it up. So antistars would shine just as stars do. The same goes for antiplanets, and antipeople. The antimoon above an anti-Earth would be made for antilovers.

It is in precisely this sense that we distinguish "nothing" from "something." If the universe contained equal amounts of matter and antimatter mixed together, it might as well have contained nothing. Unless something happened to change the balance, the matter and antimatter would have annihilated each other, leaving nothing but pure radiation. And a universe of pure radiation cannot form galaxies, and stars, much less planets, people, or atoms.

So the lives of our atom truly began at the moment when the amount of matter and the amount of antimatter in the universe started to differ. Only then could any history worth writing begin. And of course the central question that then arises is: Was this difference written in at the beginning, as on some cosmic tablet, or did it occur by accident?

Fortune favors the prepared mind. The notion that our very existence might depend on such a subtle event is not one that immediately comes to mind as you begin to think about creation. Until about 30 years ago this issue wasn't even raised by scientists because there was no scientific context in which to frame it. A serendipitous observation in New Jersey changed all that.

In 1965 two physicists at Bell Laboratories in Holmdel, New Jersey, detected an unanticipated static in a sensitive radio receiver they had tuned up in order to listen for radio signals from the sky. This static turned out to come from a uniform background of radiation bombarding us from all directions in the sky, whose source was none other than the Big Bang itself.

This cosmic background radiation (CBR) has been streaming through the universe largely unimpeded for billions and billions of years. The density of the universe was last sufficiently large so that this radiation regularly interacted with matter when the universe was 1,000 times smaller, and had an average temperature of about 3,000 degrees Celsius.

Although the CBR had many of its presently observable features imprinted at that time, when the universe was about 300,000 years old, the origins of this radiation background are as old as the universe itself. And this background has one striking feature that colors the entire character of our universe. Like all electromagnetic radiation, this radiation bath is made up at a fundamental level of individual particles, or quanta, called photons. Photons have no rest mass, and thus travel at the speed of light, a characteristic of all radiation. When we add up the number of photons in the CBR and compare that number to the total number of protons and neutrons in all the atoms in all the stars and galaxies in the universe, we find about 1 billion photons for each particle of matter in the universe today.

We happen to live in one of those rare parts of space that has lots of matter. Just as a fish might look around its immediate environment and conclude that the universe is made of water, we intuitively sense that our peculiar circumstances are generic. They aren't. Most of space is almost devoid of matter, but the radiation bath is everywhere.

Where did all this radiation come from? I have already hinted at the answer. If there had been no excess of matter over antimatter early on, radiation (that is, photons) would be *all* that would be left in the universe today. Instead, sprinkled amid this radiation is all the matter that makes up the visible universe. So the ratio of 1 to 1 billion, protons to photons, in the visible universe at the present time can tell us, indirectly, something important about the early universe.

Each particle–antiparticle annihilation in the primordial universe would have produced, on average, 2 equally energetic photons. The fact that there are about 1 billion photons in the CBR today for each proton left in the universe tells us that for every particle of matter that survived to the present era, around 1 billion particles and antiparticles in the early universe must have died trying!

Each atom today is therefore a survivor of incredible odds. In the turbulent soup that was the primordial universe, there must have been *almost exactly* as many particles as antiparticles, with just a few extra particles left over. Were it not for a small pollution at the part-per-billion level — far smaller than the detectable level of many radioactive elements in the materials that surround us today — no atoms would now exist in the universe.

Think about it! We look around the universe today and see only matter — stars and galaxies — and yet we deduce that this universe must have arisen from one where the number of particles of matter and the number of particles of antimatter differed by less than 1 part in 1 billion.

To present the peculiarity of this situation a little more intuitively, we return to our incredibly dense and hot early baseball. If it were a real baseball, we might choose to examine it under a microscope, where we could see the small strands of thread used in the stitching holding the leather outer pieces together. If this baseball were an impressionist's representation for our observable universe as it was near the beginning of time, and if we counted particles at that time, then all of what now makes up everything we see — people, planets, stars, galaxies — could have been fully contained in a single speck on a single thread. Remove that thread, and all that would have been left today is the invisible radiation bath that still surrounds us.

Realizing that the very existence of life in the universe today hung at that time at least metaphorically by the narrowest of threads, one's first reaction is to wonder why. Why was the asymmetry between matter and antimatter so insignificant? Once again, it turns out that our natural predisposition misses the point. The real surprise is that there was any asymmetry at all.

Pure energy is "antiblind." That is, Einstein's famous relation between mass and energy, $E = mc^2$, doesn't specify whether the mass is in matter or antimatter. Since antiparticles have exactly the same mass as their particle partners, given the right amount of energy it should be just as easy to convert this energy into the mass of one as the other. So among the ejecta of billions of collisions occurring every instant in the very early universe, each with enough energy to create many more new particles and antiparticles, equal numbers of particles and antiparticles should have been spewed.

But there is a roadblock to this process. In the words of Ian Fleming, creator of superspy James Bond, paraphrased later by Nobel laureate Sheldon Glashow, "Diamonds are forever." That is to say, "matter" (as opposed to mass, in a way that will become clear in a moment) apparently does not spontaneously come into, or go out of, existence. We can dilute matter by an arbitrary amount of radiation, or by an arbitrary number of pairs of particles and antiparticles, but in the world of our experience we can never get rid of it completely, nor can we create matter without antimatter from nothing.

Even before Einstein showed that mass and energy could be interconverted, chemists had discovered that chemical reactions never change the total electric charge of the reactants. Two hydrogen atoms stripped of their electrons, thus becoming positively charged "ions," might combine with 1 doubly charged negative oxygen ion to create a neutral water molecule. Positive sodium ions could combine with negative chlorine ions to produce table salt, and so on. What became known as *the conservation of electric charge* formed a central feature of the laws governing electricity and magnetism. And Einstein based his discovery of relativity on these laws, so that theory certainly didn't circumvent them. Thus when energy is converted into mass, the total electric charge produced must be precisely zero: photons, which have zero charge, convert into particle–antiparticle pairs, for example, and not to particle–particle pairs. A century of careful experimentation has only served to confirm that if you start out with no net electric charge in a system, nothing you can do, in heaven and earth, can create a net charge.

It turns out that there is a beautiful theoretical underpinning that explains why charge is conserved in electromagnetism, and it also explains why photons, alone among all particles we know of, must have absolutely zero mass. This is based on a hidden symmetry of nature, unveiled in the early part of this century. It is called a *gauge symmetry*, after the name coined by the German mathematical physicist Hermann Weyl, who first explored its mathematical details in an early, unsuccessful effort to relate the forces of electromagnetism and gravity. Although Weyl's effort was not successful, the mathematics of gauge symmetry now forms the basis of our understanding of every one of the four known forces in nature: the two familiar long-range forces, electromagnetism and gravitation, and the two short-range forces that operate on nuclear scales, called the weak and strong interactions.

Charge conservation alone cannot explain the stability of matter, however. A proton, which is positively charged, is not forbidden by these arguments from decaying into the antiparticle of an electron (a positron), plus some neutral particle such as a photon. Since the proton weighs 2,000 times as much as a positron, if there were not some powerful road-block that forbade such a transformation it would happen in an instant. Before you could say "Rumpelstiltskin," all the protons in the universe would be gone.

One of the basic building blocks of matter *is* unstable. Neutrons, the nuclear partners of protons, weigh just a tiny bit more than protons. The difference between the mass of a neutron and a proton is less than 1 part in 1,000. That difference, however, is sublime. Without it, life would not be possible. That's the good news. The bad news is that because of this mi-nuscule mass difference, neutrons can decay. A free neutron decays into a proton (positively charged) plus an electron (negatively charged) plus an antineutrino (neutral) with a lifetime of about 10 minutes, on average.

I remember I was shocked when I first learned that one of the funda-mental components of atoms was unstable. How could a fundamental part of me and you be radioactive? If a neutron could decay, how could any matter survive? The answer lies in what appears to be another mirac-ulous accident — one that will completely govern the later life of our atom. As I noted earlier, the mass difference between a proton and a neutron is very, very small, about 1/1000 the mass of the particles them-selves. Thus a free neutron is only just a bit heavier than the sum of the masses of a proton, an electron, and an antineutrino, and it is thus just only barely able to decay into these particles. Most elementary particles that are unstable have lifetimes of millionths or billionths of a second at best. Free neutrons, however, live about 10 minutes before decaying. When a neutron is located inside an atomic nucleus, it is bound to its other nuclear partners, protons and neutrons. Being "bound" in physics means it would take energy to pull the particles apart. Thus the neutron loses energy when it gets bound inside a nucleus. It turns out that the binding energy between the neutrons and protons in the nucleus usually exceeds the very small mass–energy difference between a free neutron and proton. Thus inside such a nucleus the neutron is effectively lighter than it is in empty space, and there simply is not sufficient energy avail-able to create a proton, electron, and antineutrino were it to decay. Thus atomic nuclei remain stable by an energetic accident.

Now, since free neutrons can decay into protons, they both carry some similar property of "matter-ness." Protons have no energetic barrier for decay into much lighter positrons and photons, so the matter-ness property both protons and neutrons possess must also therefore prevent the proton from decaying into anything lighter. Sticklers may of course point out that protons and neutrons are made of the more fundamental particles called quarks. But this just begs the question. What stops the quarks that make up protons and neutrons from decaying into lighter nonquarklike particles?

As far as we can tell, the stability of protons (and their constituent quarks) is, within the context of what has become known as the standard model of elementary particles, a complete accident. No fundamental property of this model keeps it stable. It just happens that there are no interactions within the standard model that would cause it to decay. Electromagnetism allows quarks inside protons and neutrons to interact with light, but not to otherwise alter their identity. The weak interaction allows quarks to interconvert their identity but in a way that allows only protons to convert into neutrons and vice versa. The strong interaction ensures that quarks are stably bound inside the proton. And gravity interacts with all matter in an identical way, and does not cause matter to decay. But this does not constrain as yet undiscovered physical processes that may exist beyond the sensitivity of current experiments. Because we are around today, however, if protons are unstable they must live orders and orders of magnitude longer than any other unstable particles we know of.

One of the first convincing proofs that the lifetime of protons must be at least much much longer than the age of our current universe was provided by the ingenious experimental physicist Maurice Goldhaber, who from 1961 to 1973 was the director of the Brookhaven National Laboratory on Long Island. For this particular demonstration, however, Goldhaber needed no experimental equipment. He wrote a paper in 1954 whose central claim was that if protons lived less than 10^{16} years, a million times as long as the present age of the universe, we would, in his words, "feel it in our bones!"

By this statement he meant the following. If protons decay into lighter particles, such as positrons or photons, the energy released would be about 1,000 times greater than that released in normal radioactivity, where a nucleus of one type converts to another type, and releases energy in the process. If protons decayed in your body, they would have a far more devastating effect than other types of radioactivity. Because there are so many protons in your body (more than 10^{27}), if each proton lived, on average, 10^{16} years (or about 10^{23} seconds), there would be more than 10,000 protons decaying in your body every second, on average. This level of radiation is likely to be prohibitively large. The fact that anyone survives past infanthood is therefore proof that protons live longer than this.

Now, physics is a two-way street. Whatever it is that stops protons from easily decaying into lighter particles must also forbid the inverse process of proton creation through the collisions of the lighter particles. If one produces enough energy to create a proton, one must create along with it the correct number of antiparticles so that the total matter-ness of the products is the same afterward as it was before.

Which brings us back again to the earliest moments of atomic conception. In this inferno, particles were being bombarded by radiation energy so often and so intensely that one could in principle create all the matter in the universe in less than a billionth of a billionth of a second. That is, if one only could! However, if one could, one could just as easily also destroy all the matter in the universe, and just as quickly!

So there's the catch. Matter cannot be created from nothing, apparently, but even if it could, the reciprocity of the laws of physics implies that it could disappear back into nothing. How can we ever hope to understand why atoms exist in this case?

One easy way out is to simply say "In the beginning, God created matter." If matter is really immutable, then if God created it at the beginning no mere physical process could get rid of it in the 12 billion–odd years since creation.

But it would be quite remarkable if one had to invoke God to explain the origin of atoms, because thus far we have been able to describe the evolution of our universe and everything inside it, back to at least the first second of the Big Bang, using only a few simple laws of physics. Moreover, a universe created with 1 extra particle of matter for every billion particle-antiparticle pairs seems somewhat awkward at best. In human history, such a ratio has never taken on any particular divine significance, for example.

On the other hand, in the history of mathematics certain numbers have carried a special significance. Zero is such a number. Indeed, the concept of zero was so powerful that those early mathematicians who first discovered it kept it a closely guarded secret! Another special number is 1. If the fractional ratio of two fundamental quantities in the early universe turned out to be zero or unity, then we might ascribe it some special significance. A ratio of 1 to 1 billion, i.e., 10^{-9}, however, doesn't seem special at all. Moreover, for those who like to believe that somehow the universe was created so that we might enjoy it, there is nothing in this ratio that seems to uniquely provide for the future existence of humanity. It is true that if the fractional excess of particles over antiparticles were zero, life could not exist. But if it were 10 or 100 times larger, for example, there is nothing I know of that would have gotten in the way of our eventual arrival on the scene.

Einstein argued, metaphorically I expect, that God does not play dice with the universe. By the same token, it seems less than satisfying to imagine that the proton-to-photon ratio, so essential in coloring the nature of our existence, was randomly chosen by a gambling God. Indeed, as I have alluded to above, if a divine being wanted to create a mathematically beautiful universe the obvious number to begin with is zero. If there were no asymmetry between matter and antimatter at the beginning, nature would be as simple as it could be. There would have been no loss of innocence, and the universe would be a peaceful, if lonely, place.

One can argue all day about which initial configuration is more beautiful than another, or how many angels can dance on the head of a pin, but such metaphysical debates usually lead nowhere. On the other hand, there is no denying that a universe with equal amounts of matter and antimatter is more mathematically symmetrical than any other initial condition. And since mathematics is the language of science, if not metaphysics, and since science seems to do a wonderful job of describing the physical universe, this special configuration does hold special interest for physicists.

Whatever one's mathematical or theological bent, however, thankfully there now seems to be no need for either metaphysics or appeals to mathematical elegance to resolve the issue of the origin of matter. In the past 30 years, developments in the physics of elementary particles have pointed to a natural mechanism for starting with nothing and ending up with something — in particular with 1 part in a billion of something. What's more, this mechanism could preserve the long-term stability of matter today. I think it is fair to say that this is one of the great, largely unheralded, surprises in modern physics. And without it, our atom is literally nowhere.

3.
TIME'S
ARROW

The thrill of unexpected discovery . . .
cannot help but stir the blood!

ISAAC ASIMOV

Near the center of Moscow stands an impressive set of buildings housing the Physical Institute of the former Soviet Academy of Sciences. Here, shortly after the end of the Second World War, a young graduate student, Andrei Dmitrievich Sakharov, began to work under the supervision of the renowned physicist Igor Tamm on the problem of how to produce the first thermonuclear explosion on Earth. Within two years Sakharov was directing the Soviet government's concerted effort to become a nuclear superpower. A continent away, the Hungarian expatriate physicist Edward Teller was similarly promoting a program in the United States to develop "the Super-Bomb," as the hydrogen bomb was then called. The careers of these two eminent physicists had remarkable parallels, and divergences. Teller, more than any other physicist in the United States, would become associated with the relentless drive toward nuclear proliferation and weapons research. Sakharov would win the Nobel Prize in 1975, not for physics but for peace, as a result of his own efforts to end the construction of nuclear weapons and to push for worldwide nuclear disarmament. His exile to the city of Gorky in 1980 for his activities provoked an international outcry, and he became a hero to a generation longing for an end to the cold war. In the end he outlived the harsh Soviet system that had exiled him.

Both Sakharov and Teller were more than mere weapons physicists. Teller contributed important ideas to nuclear physics, and to the theory of stellar evolution. Sakharov worked on a broad variety of problems, spanning many areas of physics, in the typical Soviet tradition. Following the example of the cosmologist Yakov Zeldovich, his colleague in the development of the Soviet hydrogen bomb, he even turned his attention to cosmology.

In 1967, barely two years after the discovery of cosmic background radiation, Sakharov wrote a paper of fundamental importance for cosmology, although it was basically ignored for almost a decade, primarily because his ideas were far ahead of their time. Undaunted by the fragmentary knowledge then available about the interactions of elementary particles, or perhaps oblivious to this handicap, Sakharov asked the prescient question: How could the universe generate a matter–antimatter asymmetry if none existed at the beginning?

In order to address this question, we must first remind ourselves that when we talk about a matter–antimatter asymmetry in the universe, we are really concentrating on an asymmetry between the fundamental particles making up the bulk of visible matter, protons and neutrons, and their antiparticles. Protons and neutrons are called *baryons*. If we are going to change the number of baryons in the universe compared to the number of antibaryons in order to generate an aysmmetry where none had previously existed, Sakharov recognized, as have we, that the fundamental ingredient in this process must be some new set of interactions that can independently change the number of baryons in the universe. Clearly these interactions must be extremely weak today, however, or else the proton would decay in a time much shorter than experimental constraints allow.

But perhaps more important, Sakharov determined that two other subtle conditions must also exist in order to generate an asymmetry of matter and antimatter in an expanding universe.

The first of these is a departure from *thermal equilibrium*. A system in thermal equilibrium is one in which the available heat energy is partitioned uniformly among all parts of the system. Thus, for example, when the air in this room is in thermal equilibrium, I would expect it to be the same temperature throughout. If one part of the room started out hotter than the other part, I would expect that, given sufficient time, the motion and collisions of the air molecules would eventually even things out. Similarly, when I pour milk into my coffee and stir it, I expect that eventually the liquid will become uniform in color.

Thermal equilibrium implies that if there is enough energy in a hot bath of radiation (such as existed in our primordial baseball) for collisions to create new particles, then all particles having precisely identical masses will be created in equal abundance. But the fact that particles and antiparticles have the same mass implies that in thermal equilibrium any new interaction that changes the number of baryons and antibaryons will nevertheless ensure that they will achieve equal abundance. Thus without a departure from thermal equilibrium, no matter–antimatter asymmetry can result in nature.

Finally, there is one far more subtle and strange requirement for the birth of atoms, which Sakharov, to his great credit, recognized. Here he was undoubtedly influenced by a surprising and completely unexpected discovery made three years earlier, in 1964, which would later garner the Nobel Prize in physics for the scientists involved. Sakharov recognized that in order for the universe to produce an asymmetry in matter and antimatter, time must have an arrow.

The argument is deceptively simple, if not at all obvious. Say you make a movie showing a positive charge moving to the right. If you reverse the direction of the film through the projector (thus allowing time to run backward on the screen), the positive charge will move backward, that is, to the left. If we concentrate on the flow of the electric charge during the forward run of the film, the right-hand side of the screen would get more positive as the positive charge moves toward the right. If a negative charge moved toward the right, then that side of the screen would instead get more negative. But if you run the film of a negative charge backward, so that the negative charge moves toward the left, then the right-hand side of the screen will now get more positive, just as in the original case. Thus, from the point of view of charge flow, a positive charge moving forward in time can be equivalent to a negative charge moving backward in time.

This equivalence between processes involving positive charges and the time-reversed processes involving negative charges has a powerful implication. If the laws of nature at a fundamental level are insensitive to time's arrow, then every process that can occur involving particles can also occur at precisely the same rates if all particles are replaced by their antiparticles with opposite electric charge. If neutrons can decay into protons, electrons, and antineutrinos, then antineutrons must decay at precisely the same rate into antiprotons, positrons, and neutrinos.

This in turn implies something very strange indeed. If a particle–antiparticle asymmetry developed dynamically in the universe where none initially existed, then some physical reactions must have occurred at a different rate for particles and antiparticles. But if this is true, our argument above implies that whatever force is responsible for these reactions *must* distinguish an arrow of time. Namely, it must predict different rates for identical reactions if time were run backward.

This may not seem that strange. After all, who has not watched a video, or movie, in reverse, and seen how ridiculous the action seems, with broken shards of glass suddenly coming together to form a beer bottle, or a windshield. Everything about our experience distinguishes the future from the past. One does not regret yet unperformed mistakes. And one is never hopeful that the past can get better. An arrow of time is a central feature of our everyday experience.

Intuitively, however, we recognize that the apparent arrow of time seems to result from the great complexity of nature. Broken down to a fundamental level, the underlying classical laws of motion do not appear to distinguish future from past. For example, if I film a single billiard ball bouncing around the sides of a billiard table without pockets, I can run this film in reverse and the motion will not look strange. Indeed, anyone viewing the film would not know whether it was running forward or backward. On the other hand, if I have 15 billiard balls arranged by a rack into a triangular shape, and I hit them with a cue ball, when I run the film of this event backward it looks completely ridiculous, with order apparently arising spontaneously out of randomness.

Somehow the collective behavior of a set of billiard balls is different from the behavior of a single ball. How individual particles when in huge numbers combine to produce a world where the future can be distinguished from the past is a subject rich in its own complexity and history. The first person to seriously attempt to understand it quantitatively was Ludwig Boltzmann, who, as I noted earlier, was one of the first modern scientists to take the reality of individual atoms seriously. His efforts in this regard were so subtle that it was some time before other physicists truly understood them.

But as fascinating as the history of what has become known as statistical mechanics is, I will not focus here on such a complicated subject. Instead, I wish to address the possibility that a single interaction, involving only a few fundamental particles, can exist which would produce different effects if the arrow of time ran backward. This is like requiring a single billiard ball to return to a place on the table different from where it began if one ran a movie of its motion backward.

Pool players can feel safe, however. The laws of classical mechanics guarantee that if I run such a movie backward, the billiard ball will end up precisely where it started. Indeed, all the laws of physics that operate on human scales, including both gravity and electromagnetism, have such sensible properties.

Thus for many years all physicists took for granted "time reversal symmetry" as one of several fundamental symmetries of the laws of nature. For example, no one imagined that the laws of physics distinguish right from left. If a baseball player hits a ball toward right field on a day with no wind, it can be expected to travel an identical distance were the player to have hit the ball to left field.

Our smug certainty began to crumble in the 1950s, however. On microscopic scales, nature proved far more subtle than we had ever imagined.

As I have described, the neutron is an unstable particle, and it decays, via the weak force, into a proton, an electron, and a neutrino. Neutrons are also particles that have what is called *spin* — that is, they behave as if they are spinning around some axis. In 1956 the decays of many neutrons were carefully observed by a number of groups. The direction into which the electrons were emitted, relative to the axis around which the decaying neutrons were spinning, was not distributed uniformly between hemispheres as one would sensibly expect if the laws of nature at this scale did not differentiate between left and right. In this way it was discovered that somehow the weak force does distinguish left from right.

No sooner was this result confirmed than a similar set of experiments demonstrated that the decays of another particle called the *pion*, also governed by the weak force, violated not only left–right symmetry (called *parity*) but also the apparent symmetry between particles and antiparticles in nature. The configurations of the particles resulting from the decays of pions and antipions are not identical if one merely replaces each particle with its antiparticle.

In this sense, it may look like I misled you earlier when I said an anti-world would appear to be identical to our world, since this reaction, at least, is not identical if all particles are replaced by antiparticles. However, this peculiar behavior appears confined to those reactions mediated by the *weak* force. But it is the *electromagnetic* force that governs essentially all phenomena that are observed in everyday life on human scales. Thus, in this sense, an antimatter world is essentially the same as a matter world, so strictly speaking I was not lying. I am not sure how this would hold up in a grand jury proceeding. That would depend on what the meaning of *is* is.

In any case, it was learned, almost a decade after the observations of the decays of neutrons and pions that demonstrated that the weak force distinguished left from right and particles from antiparticles, that it did something which was even stranger. I have already described that many elementary particles, such as protons and neutrons at the heart of ordinary matter, are themselves made of elementary objects called quarks. There are 6 known types of quarks, even though only 2 different quarks are responsible for the bulk of the properties of protons and neutrons. The others seem to exist to make the universe more interesting.

The first of the new quarks to be proposed and then discovered was named the *strange quark*, by the American physicist Murray Gell-Mann, the father of quarks. Gell-Mann is, among his many other talents, a linguist, and his choice of the word *strange* to describe these new objects could not have been more appropriate. Experiments in the 1960s on a new type of elementary particle called a *kaon* containing a strange quark led to a startling discovery. Using the known properties of special relativity and quantum mechanics, very careful measurements of the decays of kaons demonstrated that the reverse processes (that is, the creation of kaons in collisions of the decay product particles) would not occur at the same rate if the arrow of time were reversed! An arrow of time, at least in one very special system, had finally been discovered.

Thus three years before Sakharov wrote down his conditions for the universe to dynamically generate a matter–antimatter asymmetry, the third and most exotic of these conditions — violation of what has become known as time reversal symmetry — was discovered to actually exist, albeit in a rare and special corner of nature. At that time, however, there was no evidence for the existence of Sakharov's other two fundamental conditions: processes that independently change the number of baryons and antibaryons, or produce a departure from thermal equilibrium in the early universe.

It would be nice to end this mini-saga with the story of how particle physicists, emboldened by these three requirements, set forth to develop the necessary theoretical infrastructure to explain how the universe ended up being dominated by atoms of matter and not antimatter today. But that is not how it happened at all. Sakharov's ideas languished, and physicists went about their business of trying to explain what at the time was a bewildering menagerie of data on fundamental interactions.

And although the Nobel committee has not yet fully acknowledged it, the 1970s was perhaps the most successful decade in the twentieth century in terms of revolutionizing our theoretical picture of fundamental forces. In 1967 we understood the basic framework of only two of the four forces in nature, gravity and electromagnetism, and the zoo of elementary particles appeared to be growing without limit. By 1978, we had gained a solid theoretical foothold to describe each of the known forces, and we appeared to have uncovered the essential cadre of elementary particles associated with all the essential physical processes we observe in the universe today.

More surprising, perhaps, by 1976 all of Sakharov's ingredients had become part of the regular recipe of elementary-particle theory. And it was not long before elementary-particle theorists dusted off the papers of a decade before and realized that Sakharov's holy grail was in sight. Moreover, by exploring the interactions between subatomic particles in terrestrial accelerators, they had come to the threshold of being able to calculate, from first principles, precisely how our oxygen atom could come to exist.

The action first began to heat up in 1973. That year, the physicists David Gross and Frank Wilczek at Princeton, and independently David Politzer at Harvard, made a remarkable discovery about the nature of the strong force that binds quarks together to make protons and neutrons. Because of the huge force between quarks, understanding the detailed nature of the strong interaction had, up to that point, remained largely impervious to theoretical assault. Gross, Wilczek, and Politzer discovered, however, the most amazing property of what is now understood to be the correct theory of the strong force, called *quantum chromodynamics*, by analogy to the quantum version of electromagnetism, called *quantum electrodynamics*. They demonstrated that the interaction between quarks gets weaker the closer they approach each other. On very small scales, the interaction between particles that are very close together would be weak enough to be treated on the same footing as the other, weaker forces in nature.

Two years later, Harvard physicists Howard Georgi, Helen Quinn, and Steven Weinberg pointed out another interesting fact. While the strong force gets weaker with decreasing distance, the electromagnetic force and the newly understood weak force get stronger with decreasing distance. As these physicists demonstrated, at a very small distance scale, a million billion times smaller than the size of a proton, the magnitudes of all these forces could converge. Perhaps at some fundamental scale all the forces could be unified.

At around the same time as these ideas were floating around, Sheldon Glashow and his colleague Howard Georgi made a bold proposal. They showed that the newly understood theories of the strong force between quarks and the weak force that governs the decay of neutrons (the theory of which Glashow and Steven Weinberg had helped develop in the 1960s, and for which they would share the Nobel Prize in 1979 with Pakistan's Abdus Salam), could be combined with the electromagnetic force into a simple mathematical framework. Thus, mathematically at least, the three different forces could be viewed as different manifestations of a single underlying force operating at the extremely small distance scale characterized by the scale at which the strength of these three forces seemed to become equal. Moreover, this idea could also resolve several longstanding puzzles in particle physics, including why all elementary particles have electric charges that are integer multiples of the charge on the electron. They called the resulting theory a *grand unified theory*, or as it has affectionately become known since, a GUT.

Suddenly the smell of supreme synthesis was burning in the air! The excitement in the particle physics community at that time cannot be overstated. A giant step along the road to Einstein's goal of a single theory unifying all the forces in nature seemed to have been achieved. Within five years, we had proceeded from muddy waters to a possible utterly clear Theory of Almost Everything! All the indirect evidence pointed consistently in the same direction. Particle physics seemed on the threshold of an almost complete description of nature on fundamental scales and, as we shall see, a new understanding of why we live in a universe full of matter today.

There remained some slight problems, however. First and foremost, the scale at which the forces might be unified was 15 orders of magnitude smaller than the smallest scale that could then be directly probed by particle accelerators. Experimentalists could complain that this model might be beautiful, but it did not appear to be testable, at least directly. No accelerator in the foreseeable future will ever create the energies necessary to explore nature on such a small scale. Theorists in turn recognized the great leap of faith involved. After all, whenever we have probed the structure of matter on smaller scales, we have discovered something new and unexpected. To extrapolate one's theories to a scale 15 orders of magnitude smaller than the smallest scale on which we had direct data, and to expect to be correct, seemed to strain the limits of conceit.

However, grand unified theories unify not merely the forces in nature but also the particles of matter. In such models, quarks, the building blocks of nuclear matter, plus electrons and neutrinos are combined into a larger family of matter, along with much heavier brothers and sisters that would not be observable today, having decayed in the early universe. But this unification comes at a cost. When the strong, weak, and electromagnetic forces are combined at the GUT scale, new interactions can not only change one type of quark into another, as the weak force does when it causes neutrons to decay, they can also change quarks into particles such as electrons and neutrinos.

But if quarks can turn into electrons and neutrinos when they collide, the protons that are made of quarks will disappear when the 2 quarks inside the proton so interact. In this way, the number of particles making up observable "matter" in the universe can change as a result of these extra interactions.

If such extra interactions can cause protons to disappear, how come the stuff we are made of appears to be so long-lived? It turns out that because these forces operate at such very small scales, the quarks inside of protons essentially never get close enough, on normal timescales, to feel such forces, and thus they do not get converted. In the very early universe, however, when matter was compressed to incredible densities, the interparticle spacing would have been so small that these new interactions could take place with impunity. A means to change the number of quarks, and thus protons in the universe, then became possible.

It was a matter of months before physicists proved, within the context of GUTs, that not just the spontaneous appearance or disappearance of quarks was theoretically possible, but that all the conditions necessary for the creation of a matter–antimatter asymmetry could have existed in the very early universe.

So finally, here is the picture of our atom's birth, as seen through the filter of a Grand Unified Theory:

In the incredibly compact primordial baseball, exotic elementary particles were being bombarded by energetic radiation at unbelievable rates. At that time, in a billionth of a billionth of a billionth of a second, more energy flowed through a region that would today encompass the size of an atom than has been produced by all the stars in our galaxy in its lifetime. As a result, particles could change identity billions of times each billionth of a second. All particles that could be created were created. As long as the temperature was high enough, as many superheavy exotic particles were produced out of this radiation bath as decayed into it. However, as the universe cooled slightly, the decays got the upper hand. There was not quite enough energy to continue to produce the superheavy particles, and they began to disappear.

These doomed particles — call them X-particles — could perhaps decay into 2 quarks, or perhaps into a quark and an electron. Since both decays produced quarks where none existed before, baryon number, a.k.a. "matter-ness," was produced. Remember, however, that for every X-particle around at the time, there was a corresponding X-antiparticle. The decays of the X-antiparticles then produced either 2 antiquarks, or an antiquark and a positron (the antiparticles of the particles produced in the X decays).

If the X-antiparticles were to decay into the antiparticles of the particles produced by the decays of the X-particles at precisely the same rates, then as many antiquarks would be produced as quarks. But if there were even the tiniest difference in the rates of decay for each of the two decay channels of X-antiparticles, compared to X-particles, then the total number of quarks left over after all the X-particles and X-antiparticles decayed might have been ever so slightly different from the number of antiquarks.

A small difference is all that was needed. One extra quark produced in the early universe for every 1 billion quarks and antiquarks would be sufficient to account for all of the matter we observe in the universe today. In the course of time, the rest of the quarks and antiquarks would have been annihilated, producing the roughly 1 billion photons in the cosmic radiation background for every proton in the universe that we observe today.

A very small step for the universe, but a huge leap for mankind! For in this imperceptible yet immutable difference lay the seeds for all the atoms and all the people and all the stars and all the galaxies in the universe today.

Once described, it doesn't sound like much. Who would have believed that such an inconsequential, minute, pitiful inequity would have such remarkable consequences? And who would have believed that it would take Newton, Maxwell, Boltzmann, Einstein, Dirac, Heisenberg, and the rest to uncover this possible hidden blemish in the cosmos? If this is God's hand in creation, it is the smallest hand one can possibly imagine. But theological speculation aside, this *was* the defining moment in the creation of the universe of our experience. While that 1 extra quark would, in a fraction of a second, undergo myriad collisions, interactions, and transformations following its production, its descendants would never, ever again lose their essential "quarkness." Nothing in the future of the universe up through the present time could remove this tiny extra bit of matter. What the X-particles had once produced, no man, and no collision, could put asunder.

Once created, the quark's matter-ness would flow through a river of time. For each and every atom in your body, there is a set of quarks, created in those early instants, to which we could trace your existence if we had the computational means to do so. The atom of oxygen I breathe now, which helps to give me the energy to tap the key to type this word, is as connected to one specific set of quarks created in the primordial baseball as I am to my great-grandfather's great-grandfather. Perhaps more so.

Of course, tracing my oxygen atom back to the beginning of time, to the anonymous decays of some unknown X-particle, is only remarkable if it is correct! And alas, we don't yet know whether this picture, plausible and succinct as it is, has anything to do with reality. Debating about how many particles and antiparticles were produced eons ago may seem to be speculation in the extreme. Hypothetical X-particles last existing at the moment of creation may seem no more real than hypothetical *X-Files* do today! But good physics is not based on hindsight, or fantastic storytelling. To be science, GUT models must be testable.

Which brings us back to today, to the dark mineshaft in Japan, and a large tank of pure water. Nature is subtle, and part of the wonder of science is to seek out the subtleties. Protons, containing quarks alone, and residing in our calm, old universe, may seem immune to the vicissitudes of X-particle creation and decay. But they aren't. One of the most striking predictions of Grand Unified Theories is that ultimately, even diamonds are not forever.

I alluded to this result above. Quarks inside of protons *almost never* get close enough together to feel the grand unified forces, but *almost never* and *never* are not the same. If one waits long enough, 2 quarks inside of a proton, inside of an oxygen nucleus, inside of a water molecule, inside of the Super-Kamiokande tank, inside of the Mozumi mine, are destined eventually to brush close enough together and change into other particles such as antiquarks and positrons, causing the proton they are a part of to disappear.

The fact that Grand Unified Theories require the collisions of the quarks inside protons to momentarily produce and exchange X-particles that may be billions of billions of times heavier than the proton in which the whole process takes place is not an insurmountable problem. Quantum mechanics tells us that as long as these superheavy particles are exchanged over such a short time that you cannot measure their presence directly, even in principle, then the fact that there doesn't seem to be enough energy around to create them in the first place isn't a problem. Like it or not, quantum mechanics shares a common thread with the actions of various U.S. presidents: If you don't get caught, you didn't do anything wrong! It may not seem fair, but it's the way the world works.

Now, the probability that 2 quarks inside a proton will get close enough together to go *poof* in this way is very small indeed. One can calculate that the mean lifetime for protons to disappear via such a process is at least a million million billion billion years. This is over a hundred billion billion times longer than the current age of the universe! It doesn't seem necessary to rush out and sell your Microsoft stock.

More to the point, perhaps, it may appear that this particular GUT prediction is also not amenable to direct verification. Experimental physicists are crafty, however, and the laws of probability are sublime. If the mean predicted lifetime of a proton is 10^{30} years, then the probability of any *particular* proton decaying *this year* is 1 in 10^{30}, an infinitesimally small number. However, if you start out with 10^{30} protons, then on average you would expect 1 of them to decay within a year.

Now, where do you get 10^{30} protons in one place? Water, of course. H_2O contains 2 hydrogen atoms for each oxygen atom. But hydrogen is nothing other than a proton surrounded by an electron, and each oxygen atom contains 8 protons. In 1 cubic centimeter of water there are approximately 10^{23} protons! This is a lot of protons, but it is still 10 million times too few to do the job. One therefore needs at least 10 million cubic centimeters of water. This would be a tank of water 3 meters on a side. To be safe, and to get more than one event in the lifetime of the experiment, you build a tank at least 10 meters on a side.

What signal do you search for? Well, in the proton-decay discussed earlier, 2 quarks convert into an antiquark and a positron. The latter will shoot out of the decaying proton. When a charged particle travels through water with a very high energy, it emits a burst of light, just as a supersonic jet emits a sonic boom. One can then place sensitive light detectors around the volume of water. If one buries such a tank deep underground, so deep that meddlesome cosmic rays from space cannot penetrate down to the tank, and takes water so pure that no radioactive decays occur inside it to fool the experimenters, then you have a proton-decay detector. And a single gold-plated proton-decay event would point all the way back to the origin of matter.

Shortly after Grand Unified Theories were proposed in 1974, a variety of experiments were launched to search for decaying protons, the biggest of which was conducted in a salt mine just outside Cleveland. I remember that when I was a graduate student in the early 1980s it seemed just a matter of time before the process would be seen, and my professors would pick up another Nobel Prize.

So far the news is not so good. We are still waiting, and the Cleveland detector is no more.

Indeed, by the mid-1980s large underground water experiments had ruled out the original GUT model and its predictions of proton decay. A lower limit was placed on the proton lifetime of 10^{32} years, almost 100 times longer than the original expectations. This might have been the end, for both experiment and theory, except for two bits of serendipity — one experimental, one observational — of the sort that make the progress of science so fascinating and unpredictable.

The theoretical development involved the recognition that a new symmetry of nature might lie buried deep in the standard model of particle physics. This symmetry, called *supersymmetry*, holds out great hope for explaining many aspects of nature at its fundamental scale. It does so at a cost, however. In particular, it predicts that every known particle in nature should have a *new* partner, a *sparticle*, if you will, none of which have been observed!

Of course, an optimist would instead say that half the predicted particles in supersymmetric theories *have* been observed. And we are willing to be optimists for two reasons: (1) The predicted sparticles (conveniently) all have masses too large to have yet been directly detected in accelerator experiments. (2) Supersymmetry makes GUT models sensible.

Now that we have measured the relative strengths of the strong and weak forces with good accuracy using particle accelerators, we can confirm that they cannot merge along with the electromagnetic force at a single scale as was predicted in the context of the original GUT models. When supersymmetry is added, however, the additional particles yet to be discovered change the prediction in just such a way that the three forces can merge together beautifully. This is a remarkable prediction, and one that adds credence to the GUT picture.

Moreover, the distance scale at which this merger takes place is perhaps an order of magnitude smaller than had been supposed in the initial GUT models. The implication of this for proton decay is clear. If quarks must find themselves even closer together in order to annihilate each other, the proton will live longer. Current estimates in supersymmetric models are in the range from 10^{34} to 10^{35} years, well beyond the current limits!

At the same time as these theoretical developments were under way, nature had further surprises in store. The large proton-decay detectors that were built in response to the GUT ideas turned out to be useful for something else. Because of their immense size, they were ideal detectors for neutrinos, those exotic particles produced by neutron decays and by nuclear reactions inside the sun and stars. On February 23, 1987, two large proton-decay detectors, the IMB detector in Cleveland and the original Kamiokande detector (a precurser to the current, much larger, Super-Kamiokande detector) in Japan, recorded 19 neutrino events in a 10-second interval. This does not sound like a lot, but in the neutrino business it is akin to being drowned with data! When the dust had settled, it was clear that these detectors had observed the neutrino signal from a star that had exploded more than 100,000 light-years away on the other side of the Milky Way galaxy. The field of neutrino astronomy had been inaugurated.

Once it was recognized that proton-decay detectors could do double duty as neutrino detectors, the Japanese government authorized funds to build the Super-Kamiokande detector 10 times bigger in the same mineshaft. No building project of that scale had ever been performed so deep underground. The logistics were incredible. Fifty thousand tons of water had to be purified, stored, and kept spotlessly clean for years on end. Over 11,000 sensitive state-of-the-art photomultiplier tubes had to be wired, and monitored without fail 24 hours a day, 365 days a year. But the Super-Kamiokande detector was opened on precisely the day in 1995 that had been chosen years in advance.

So here we are today, watching and waiting. The Super-K detector has the volume and sensitivity to detect the decays predicted by most supersymmetric models. Not surprisingly, it turns out that in the dozen years since the first generation of proton-decay detectors were built, ingenious theorists have proposed other exotic possibilities to explain the generation of a matter-antimatter asymmetry in the early universe, even if, heaven forbid, there is no GUT.

Nevertheless, the picture presented here is eminently plausible. There is good reason to believe that our oxygen atom can trace its ancestry right back to that primordial baseball. At that time a cosmic lottery took place. The odds against winning were a billion to 1. But lotteries usually have a winner, and this one was no exception, even if these winners would have been completely unheralded at the time. The stakes were high. To win meant bringing into existence a visible, vibrant universe of life and consciousness, even if the tangible rewards were billions of years down the road. Each and every atom in the universe began life only as a result of great fortune against tremendous odds. Of course, it is a long road from there to here, from quarks to atoms to humans, but the violent serendipity associated with the birth of matter would continue to echo throughout the lives of our atom.

And if our atom's existence began with an innocuous set of X-particle decays 12 billion years ago, it is possible that 1 proton will have decayed in the Super-Kamiokande tank by the time this book is translated into Japanese.*

---

*Unfortunately, in December 2001, a massive accident occurred at Super-Kamiokande, shattering eight thousand phototubes. It will take an estimated two to three years to rebuild the damaged detector so it is fully operational once more.

4.
NATURE
OR NURTURE?

Three quarks for Muster Mark!
JAMES JOYCE

I f you drive to eastern Long Island, off the expressway, you will pass the summer homes of the superrich and the merely rich that give this area its special character. The Hamptons, East and West, are located within a 50-minute drive of an innocuous-looking laboratory that has been the home of discoveries leading to several Nobel Prizes. Here at Brookhaven National Laboratory, for over 40 years elementary particles have been torn from matter, accelerated to high velocities, and rudely smashed against new targets, all to determine the nature of the forces that determine why the world is made of atoms. On the site of the original accelerator that started the modern revolution in our understanding of the strong force in 1974, a new accelerator has been built to attempt to re-create the quark soup that first congealed to form matter.

The new accelerator is called RHIC, relativistic heavy ion collider. Instead of accelerating fundamental elementary particles such as electrons or protons to high energies, this device smashes the nuclei of heavy atoms like iron or uranium against one another. Such nuclei contain hundreds of protons and neutrons. If the collision energy is high enough, it is possible that conditions comparable to those in the early universe can be re-created over large volumes — large, at least, compared to the size of a proton, although still microscopic in an absolute sense. If one is very lucky, one might "melt" the entire microscopic collision region, heating it to such a high temperature that within it quarks might momentarily behave like their nascent cousins did 10 billion years earlier, well before quarks combined to form the particles we see today.

Further down the east coast of the United States, a very different particle accelerator, one with beams of electrons smashing nuclear targets, is located a few miles from scenic Williamsburg, near the beautiful beaches of Virginia. Here experiments are carried out to explore how the properties of the nuclei of atoms arise from the underlying properties of the quarks that have combined to make up their protons and neutrons.

Both of these very different types of machines have been designed to address complementary aspects of the same vexing puzzle. All the matter we see in the universe arose from a small excess of quarks over antiquarks which, as I have described, presumably developed in the earliest moments of the Big Bang at incredible extremes of temperature and density. Yet the nuclei of the atoms that make up the matter we see today do not directly mirror the properties of their constituent quarks. Sometime before the universe cooled to a temperature of 10 billion degrees, by which point essentially all the protons now existing had been formed, the physics of quarks had to transform to the physics of nuclei.

By anyone's standards, 10 billion degrees is hot. But compared to those astronomically large quantities we have confronted thus far, it is manageable. One can imagine the number 10 billion. It is the amount of money Bill Gates's net worth changes by each time Microsoft increases or decreases its stock value by 10 points. The total number of people who have ever lived on Earth is perhaps 50 billion. Counting to 10 billion would take you about 100 years, if you counted as fast as you could and didn't take a break. Spending 10 billion dollars, on the other hand, takes Congress less than an afternoon.

It took the universe about 1 second to cool from the primordial baseball era of the preceding chapters to a temperature of about 10 billion degrees. This may seem like no time at all, but our internal clocks are not the appropriate ones to use here. A reasonable clock might be one attached to a particle in the radiation gas, ticking once each time the particle had an interaction with another particle. I have made an estimate of the number of collisions that would have taken place in 1 cubic centimeter of the universe between the primordial baseball era and what we might call the nuclear chemistry era, when the universe was 1 second old. The answer is *very* large or, as we scientists say, humongous. Roughly

10^{89} collisions would have occurred in the first second. For comparison, during its 5 billion years of burning, in each cubic centimeter in the fiery core of the sun a total of about 10^{55} interactions have taken place. This is about 10 million billion billion billion times fewer collisions than occurred in the same volume in the universe's first second. The number of collisions of the atoms in this volume of air during the 4-billion-year history of life on Earth is about 10^{45}, about 10 billion times smaller still!

The first second was thus a very busy one. Indeed, if in this book I devoted equal time to this era, based on activity as defined by number of interactions of the particles involved, all of cosmological history following this instant would not warrant a single comma. However, the virtue of writing a biography, even a cosmic biography, is that the author gets to choose what is significant. And in spite of the myriad changes that took place during this fantastic second, the inexorable march from an initial excess of quarks over antiquarks to a universe full of protons and neutrons was almost completely prescribed when the last X-particle decayed, at the dawn of time.

Almost.

You see, up until the universe was about 1 millionth of 1 second old, the constituents of the primordial soup still did not yet directly resemble any object we commonly see around us, or have yet isolated in our laboratories. At that time, free quarks were everywhere, and matter had no mass.

No scientist in any laboratory on Earth has ever seen a single lone quark, despite the fact that we know quarks make up the nuclei in everything around us. We now understand the reason for this. As I noted earlier, quarks exert a force on each other that gets stronger the further they separate. It would take an infinite amount of energy to pull 2 quarks infinitely apart from each other. They appear to be forever "confined" to live inside particles such as protons and neutrons of which they form the constituents.

Whenever I think of quarks inside of nuclei, I am reminded of that frustratingly wonderful novel by the Japanese author Kobo Abe, *Woman in the Dunes*. Here the unsuspecting lover of a lonely woman becomes trapped in her house, which is surrounded by sand dunes on all sides. Every time he tries to escape by climbing the dune walls, they crumble at his touch and he slides back. Quarks too may be born free, but everywhere they are in chains.

The precise details of the transformation between quarks, the ultimate building blocks of matter, and protons and neutrons, the building blocks of atoms, are still not completely worked out, but we know that the transformation itself could not have been very dramatic. When the universe was about 1 millionth of 1 second old, the average distance between quarks in the primordial soup was smaller than the average distance between quarks when they are inside of a normal-size proton or neutron. Say I draw a picture of this gas of quarks, showing where each quark is momentarily located (figure 1).

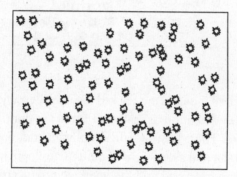

Figure 1

Now, if the average distance between quarks is smaller than it would be if each of them found themselves inside of protons or neutrons, then calling this gas a gas of protons and neutrons instead of quarks is somewhat arbitrary. How does each quark know which proton or neutron it finds itself in? For example, I can draw in lines for protons and neutrons as in figure 2.

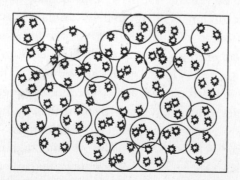

Figure 2

Or I could draw in other lines to represent protons and neutrons (figure 3).

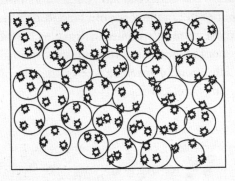

Figure 3

The point is that if quarks are packed together densely enough, what we call protons and neutrons is really as arbitrary as drawing the circles I have drawn here. There is no significant physical difference between a dense gas of quarks and a dense gas of protons and neutrons.

As the universe expanded and the quark density decreased, however, the average distance between quarks also increased. But as quarks got further and further apart, the strength of the force between them grew. Eventually, quarks became bound in well-separated protons and neutrons.

But what if, during this process, some quarks could not find nearby partners, as the quarks in the upper-left-hand side of figure 3? Well, as each quark moved farther and farther from the next nearest unbound quark, the interaction energy between these 2 quarks became so great that it began to exceed the rest mass of the quarks. Remember that it takes an infinite amount of energy to move 2 quarks infinitely far apart. In this case, it becomes energetically favorable for quarks (and anti-quarks) to pop out of empty space, as the laws of quantum mechanics allow. By appearing nearby lonely quarks and binding with them to form protons and neutrons, these new particles reduce the original quarks' interaction energies with faraway quarks by more than the energy required to make the new quarks materialize. Once again, quantum mechanics saves the day, and in the end, we finally have the building blocks of the atoms that we have known and loved for the better part of a century.

Actually, had not another, much more dramatic transformation taken place in the universe before this time, the quarks and electron building blocks would not have been able to bind together to form atoms. Sometime around a millionth of a millionth of a second after the Big Bang, we currently believe, elementary particles such as quarks suddenly became massive. Had this not occurred, the rest of history would not have been worth telling. Only because matter has mass can it collapse into the structures such as the atoms, stars, and galaxies that make up our visible universe.

I find it oddly comforting that one of the things we may take for granted more than any other, our own corporeality, was not built into nature at the beginning. If our current picture is correct, even that is an *environmental* phenomenon. For example, we on Earth are accustomed to the fact that there is a well-defined sense of "down" at any place; if you drop something, it falls down. But we also recognize that "down" is an accident of our circumstances. If we live in Australia, "down" is in a completely different direction than in New York.

"Down" is environmental because we live in the environment of the Earth. We have come to grips with this simple fact over the past five centuries, following the original Greek discovery that the Earth is round, and Newton's later discovery of the nature of gravitation. Over the past 30 years, physicists have recognized that many other aspects of the world of our experience may reflect similar environmental accidents.

The trigger for this new understanding was the realization that two of the four basic forces in nature, electromagnetism and the weak force, while totally different in character, could be identical at a microscopic level, and that the observed differences could be completely environmental in origin.

Imagine, for example, that the surface of the Earth was sandwiched between two large electrically charged spherical plates. If the outer plate had a negative charge, and the inner plate a positive charge, then any positive charge on the surface of the Earth would be attracted upward to the negative plate (opposite charges attract, remember). If the electrical charge on the outer plate were appropriately balanced, then certain positively charged particles, say protons, might be made to "levitate," with the upward electrical force balancing the downward gravitational force.

In the world of our experience, then, protons would behave as if they were weightless on the Earth's surface. Electrically neutral objects like ourselves would not feel the electrical force, and would have our normal weight. Negatively charged objects like electrons would in fact be repelled downward by the outer plate in addition to being attracted downward due to gravity, and would thus appear to weigh more than they would otherwise.

All of this might appear to be fundamental, until we discover the existence of the plates. Then we would recognize that the underlying properties of all these objects were different than we had imagined, and that what we had previously assumed to be fundamental was, in fact, an accident of our circumstances.

This is precisely what occurred in particle physics over the course of the 1960s and 1970s. The fundamental perceived differences between electromagnetism and the weak force were instead understood to be an accident of our circumstances. As a result of underlying dynamic factors that are only partially understood at present, a background "field" is believed to have developed in nature when the universe was about a millionth of a millionth of a second old. Like the electrical field in the example above, this background field affects the dynamics of certain elementary particles. The carrier of the electromagnetic force, photons, are oblivious to the presence of this field, and behave like the massless particles they actually are. The particles that convey the weak force are affected by the presence of this field, however, and they behave like massive particles in such a background. As a result, therefore, the weak force began to behave very differently from the electromagnetic force after the universe was a millionth of a millionth of a second old.

This background field, called the *Higgs field*, is not directly detectable, but its existence is inferred by its effect on elementary particles. Moreover, if this picture is correct, there are associated elementary particles, called *Higgs particles*, that must be creatable at new particle accelerators which should be operational within the next decade. Assuming this picture is indeed correct, it is precisely this field that gives mass to all the particles that make up matter. Quarks are heavier than electrons because the strength with which they interact with this background field is different. The effect is like pushing a heavy object along a smooth floor.

The object presents a certain resistance to your push, by which you infer that it weighs a certain amount. But if you suddenly encounter a rough spot on the floor, the resistance immediately increases. You have to push harder to make the object move. Again, you recognize that this effect is environmental, and is not an intrinsic property of the object. Similarly, the inertia that elementary particles display — their resistance to changes in their motion under the action of external forces — is thought to be affected by their interaction with this background field that formed when the universe was a millionth of a millionth of a second old. Without this field, matter could not exist in its present form. While massless quarks might bind together, the objects they might form would have properties vastly different from those of the particles of matter that make up our observed universe. Indeed, in such a world, stable nuclei and atoms would simply not arise.

In any case, until the universe aged by a factor of 1 million, it was not yet ready for them.

5.
TEN MINUTES
TO DIE

*The initial mystery that attends any
journey is: how did the traveller reach
his starting point in the first place?*

LOUISE BOGAN

On the Pacific atoll of Eniwetok, on November 1, 1952, at 7:14 and 59.4 seconds, it is quite possible that all the elements ever created in the universe, and some that may never have existed before, were momentarily assembled, at the same time that every living thing on the island of Elugelab was instantly vaporized. Humankind had harnessed the energy of the stars, in a bomb called Mike.

The Mike explosion on that November morning was not designed to do "pure" science, although in fact element 100, later named fermium, was first isolated from the bomb's debris. The purpose was to learn how to destroy things more efficiently, and the thermonuclear explosion that destroyed Elugelab took at most a few seconds to engulf the entire island in a fireball.

The Big Bang, on the other hand, which was also not designed to do pure science as far as we know, took several minutes to form any elements heavier than hydrogen. And even with the firepower of the biggest explosion in cosmic history, the universe then essentially stopped its nuclear cooking after the second-lightest element, helium, was produced.

Of course, the universe still wins bragging rights on almost all other counts. The maximum temperature of the Mike fireball was at most about 100 million degrees, over 100 times lower than the temperature at which the universe began building its nuclei. And the Mike explosion barely produced enough helium to fill up a large weather balloon. The universe, by contrast, in those few minutes produced over 90 percent of all the helium nuclei that now exist, accounting for about 25 percent of the visible mass in all the stars and all the galaxies our telescopes can see.

At 10 billion degrees, the 1-second-old universe was a place vastly different from the primordial inferno where quarks were born, in part due to the cosmic transformations that were described in the last chapter. Moreover, what would become the presently observable universe had by this time finally evolved to occupy an astronomical scale: about 2,000 times the size of our solar system, or about 1 light-year across, almost the distance today between ourselves and the next closest star.

The average density of the universe at this time was about 1 gram per cubic centimeter, the density of water. This is a density we need not merely imagine; we see it every day. Except for the fact that its temperature was about 10 billion degrees, it almost seems like home. Most important of all, for the very first time in cosmic history the universe's microscopic components finally began to resemble those we actually observe today. There are protons, neutrons, and electrons in your own body that date back to that moment in time.

One need not even have a machine as sophisticated as the RHIC accelerator on Long Island described earlier if all one wants to do is cook conventional nuclei. The ability to re-create the processes by which our primordial protons and neutrons combined to form the very first nuclei arose 50 years ago, with the development of the nuclear reactor, and alas, via the nuclear weapons of which Mike was the vanguard. It may not have been until the Mike explosion that atoms on Earth first re-experienced the intensity of their birth pains, but even before that awful mushroom cloud rose over the Pacific, physicists had begun to realize the alchemist's age-old dream of transmuting elements.

Sixteen very special particles, 8 protons and 8 neutrons, make up the nucleus of the oxygen atom that will be the chief protagonist of our story. Each particle is special because each was uniquely chosen from the 10^{78}

or so protons and neutrons in the observable universe to eventually form an oxygen nucleus that would, after 5 billion years or so, find itself on the surface of a green-blue planet at the outer reaches of a galaxy at the outer reaches of a cluster of galaxies that itself is on the outer reaches of a supercluster of galaxies that will one day gobble most of what we can see.

When we look around the surface of our planet, we cannot find rocks older than about 4 billion years, so we tend to think of these as the oldest objects on the surface of the Earth. This is wrong, of course. The particles that make up the rocks, even the youngest rocks that were spewed just yesterday as part of the lava flow from an active volcano in the South Pacific, are far older still. At least 7 and most likely 10 of the 16 particles in our oxygen nucleus have existed unaltered since the universe was 1 second old. Each particle was around before the first star in the universe blazed to life.

The oldest particles include at least 7 protons, and there may be 2 lucky neutrons that date back to the same era. One of the protons is a few minutes younger than the rest. As we shall see, the other neutrons are likely to be at least a million years younger, although nevertheless millions, if not billions, of years older than our Earth and sun.

When the universe was almost 1 second old, each of the 8 neutrons and 8 protons in question were changing their identity back and forth at a rate of about 10 times per second, so that to pick out which was which is difficult. As the universe expanded and cooled, however, particles continued to move farther and farther apart from their neighbors. By the time the universe had aged past 1 second, the protons and neutrons were about 100,000 times their own diameter away from each other. The abundance of other particles in the radiation gas — electrons, neutrinos, and their antiparticles — with which they could interact was much greater. But even the density of these particles had decreased substantially by this time. The average distance between a proton and the nearest electron with which it could combine to form a neutron was still 1,000 times the proton's own size. Thus by the time the universe was 1 second old, protons and neutrons could no longer pair up with electrons or neutrinos fast enough to compete with the rate at which the particles were separating. Protons, at least, were now here to stay.

Neutrons, however, have always led a much more precarious existence. They even started out with a disadvantage. Because they are ever so slightly heavier than protons, as the temperature cooled down it became energetically more favorable for the collisions that interconvert protons and neutrons to produce protons rather than neutrons. By the time the universe was 1 second old, and the proton and neutron numbers were frozen in place, there were only 20 percent as many neutrons left as protons.

(For those who like to wonder about these things, this represents another cosmic coincidence. The residual neutron-to-proton ratio differs from precisely either zero or 1 only because the average energy of the radiation gas at the temperature when the number of neutrons and protons gets frozen is very close to the difference between the rest masses of the proton and neutron. Changing the strength of the weak interaction slightly would change the freezeout temperature, and could thus dramatically alter the remnant neutron–proton ratio at that time. If the neutron–proton ratio today were zero, the sun might not yet be bright enough to support life on Earth. If it were unity, there would be no stars like the sun.)

From the moment the neutron-to-proton ratio froze in place, things got even more desperate for neutrons. Each one then had about 10 minutes left to live, all other things being equal. But another remarkable coincidence saved the day. Although neutrons and protons had stopped interconverting after 1 second, neutrons and protons could begin to bind together, thanks to the strong force, into atomic nuclei. This remained possible because the probability of a neutron and a proton binding together into the nucleus of heavy hydrogen, deuterium, is much higher than the probability of a proton capturing an electron to convert into a neutron. Once the proton and neutron were bound together, the neutron became safe. Arguments based on energetics, as I have described, would now prevent the neutron from decaying.

There remained a problem, however. The binding force between a proton and a neutron in deuterium is very weak. Thus, as long as the temperature remained high enough, energetic collisions with any particle in the radiation gas, including photons, could break apart the fragile deuterium nucleus. Several of the neutrons that would later become part of our oxygen atom may have momentarily been a part of a deuterium nucleus during this time, only to see their lifeline broken almost immediately after it had been created. By the time the universe was about 200 seconds old, however, the radiation gas had cooled sufficiently so that deuterium could now begin to form with impunity.

In this time interval, almost 40 percent of the original neutrons made during the Big Bang would not survive the wait, and would have decayed into protons. There were originally, at 1 second, about 2 neutrons for every 10 protons, so that after the neutron decays there would have been only slightly more than 1 neutron for every 10 protons in the universe. For every 10 protons around at 200 seconds, perhaps 1 of these was thus created from the decay of a neutron in the intervening period. Since there are 8 protons in our oxygen atom, there is thus a reasonable chance that 1 of the 8 originated as a neutron in this way, and is thus about 200 seconds younger than its relatives. Twelve billion years later that 200-second difference doesn't seem like a lot.

Of course, things did not stop there. Because the binding strength of neutrons and protons in the nucleus of the second-lightest element, helium, is much greater than the binding of a proton and a neutron into deuterium, shortly after the deuterium nuclei (p + n) were made the deuterium nuclei would collide with protons and neutrons to form the nucleus of helium (2p + 2n). In this way, essentially all of the surviving neutrons in the universe quickly got amalgamated into the nuclei of helium atoms in the first few minutes after the Big Bang.

There are two amazing features about this whole process. The first is that the time at which the universe first cooled sufficiently to allow deuterium to form was about 200 seconds, suspiciously close to the average lifetime of a free neutron, about 600 seconds. If the binding of protons and neutrons into deuterium had been significantly different, then the universe might have had to cool for a much longer time before deuterium formation would have been possible. But after a longer time, all the primordial neutrons would have already decayed! Helium was thus only formed in the Big Bang because the binding energy of protons and neutrons into deuterium took precisely the value it did.

The second amazing thing is that we can use this picture to automatically *predict* how much helium should have been produced in the Big Bang. Since there are 2 neutrons in a helium nucleus, if all available neutrons ended up in helium the fraction of remnant helium nuclei compared to remnant protons would be slightly bigger than about half the original ratio of neutrons to protons. (It would be *precisely* half the original ratio, except that some of the original protons would now be bound into the helium nuclei, leaving fewer remnant protons.) However, since a helium nucleus weighs about 4 times as much as a proton, the fraction of the *mass* of the universe in helium compared to that in protons would be about twice the neutron-to-proton ratio. Since this ratio was about 12 percent at the time deuterium started to form, that means we can *predict* that the primordial abundance of helium after a few minutes should have been roughly 24 percent by mass. Today, when we add up the helium inferred to be in stars, interstellar gas, and so on, the helium-to-hydrogen ratio is remarkably uniform and is between 24 and 28 percent. This agreement between the predictions of Big Bang nuclear cooking theory, based on laboratory measurements of nuclear reactions, and observation in the universe is yet another reason why we know the Big Bang really happened.

So by this time we have accounted for the birth of all 8 of the protons that would eventually help comprise our oxygen atom. They were all created within a minute or so after the Big Bang, and have been shuffling around for 12 billion–odd years to end up precisely where they are today. The history of the neutrons, however, is not so clear. One in 10 baryons that survived the first minute was a neutron, and after 3 minutes or so these were essentially all located in the nuclei of helium atoms. Remember, however, that only about 7 percent of the nuclei around at that time were helium nuclei (the rest were hydrogen nuclei, namely protons). In the intervening 12 billion years, nature has found other ways to make helium, and more complex nuclei. This is fortunate for us, because if protons and helium nuclei were the only building blocks around, life would probably never have arisen. But this means that some of the original protons in nature were processed into neutrons between the Big Bang and today. Some of the neutrons in our atom are therefore neophytes who hid out as protons until the time was right. To find out which, if any, of these latecomers actually made it all the way to our oxygen atom, we have to wait not minutes, but billions of years.

Whether originally neutrons or protons, the 16 particles in our oxygen atom today can be identified with 16 specific, but otherwise unrelated, particles in existence when the universe was a few minutes old, with nothing but destiny to later connect them. Given the intense inferno of their first seconds, followed by the desperate minutes of nuclear burning during which some particles were saved, and others forever lost, the period that followed might seem like an unbelievably long holiday.

In any case, while these particles all had a common origin, their histories now began to diverge. It was time for the universe to find proper, if temporary, homes for each of them. For these infants, nature had played its course, and now further nurture would create diversity where none had existed before. All the while, behind the scenes, events were brewing that would one day tie these particles together for a virtual eternity.

Nevertheless, at this time we must remember that not even a single atom yet existed in the universe. That is, if we mean by *atom* an object that gives elements the properties we observe of them today. The single protons and the neutrons and protons that form the nuclei of, respectively, hydrogen and helium had not yet been clothed by their electron outer blankets. All that makes the elements what they appear to be, all of chemistry and the reactions that drive life, derives from the electrons that surround nuclei. A nucleus without its electrons is like a Buckingham Palace guard without his uniform.

6

ONE HUNDRED MILLION YEARS OF SOLITUDE

The infinite quietness frightens me.

BLAISE PASCAL

The electron is the lightest particle on Earth. In mass it is essentially insignificant, accounting for less than 1/2000 the mass of everything we see. We could get rid of every electron in our bodies and we would never notice the difference if we stood on a scale. Yet, in spite of their puny heft, electrons may be the most important particles in nature, at least to us, because they determine almost every observable aspect of our existence.

The electron was the first elementary particle discovered in nature, slightly more than a hundred years ago, when Lord Thomson measured the properties of the eerie "cathode rays" observed as a glow when an electrical current passed through a vacuum tube. We now understand this glow to be associated with the light released following chemical excitations of the electrons in the few atoms of gas remaining in the tube as they are bombarded by the moving electrons in the current.

Electrons, like their baryonic cousins, protons and neutrons, have antiparticles, called positrons. When the universe was about 1 billion degrees in temperature, the numbers of electrons and positrons were almost equal. Within a few minutes, however, by the time helium had formed, electrons would have repeated the same operetta performed by protons and neutrons in the first millionth of a second after the Big

Bang. Almost all the electrons and positrons would annihilate one another, leaving only one-billionth as many electrons as had started the whole shebang, surrounded by a sea of radiation. These electrons, one for every proton in the universe, would eventually pair up with the protons to form neutral matter. But until the universe was about 300,000 years old, they were just one part of the hot gas of particles jostling each other millions of times each second.

You may wonder how we know that the number of electrons and protons in the universe is essentially identical. Well, the electromagnetic force is 40 orders of magnitude stronger than the gravitational force. If there were even 1 extra electron for every 10 billion protons on Earth, or in our galaxy, the electric repulsion would be so great that structures such as stars would not form by gravity. It is yet another curious feature of our universe that it is electrically neutral. This didn't have to be the case. One could easily imagine a universe where net electric charge was created, just as baryons were, at early times. But we probably couldn't live in such a universe.

In any case, the electrons, protons, and light nuclei created in the first minutes of the Big Bang now lay slowly cooling. The rush of creation began to subside. Physical processes slowed in inverse proportion to the age of the universe. Minutes turned to hours, hours to days, days to years, years to millennia, although these time periods did not yet have any physical representation. No planets yet existed to orbit yet nonexistent stars that might later provide cosmic clocks by which civilizations could tick off the days of their lives. For more than 100,000 years the universe would simply get colder, cooling from more than 1 billion degrees to a mild 10,000 degrees — the temperature near the surface of the sun — and virtually nothing else of significance would have been manifest.

This long period of quiescence stands in striking contrast to the rapid labor pains of our particles' birth. A story comes to mind of a young boy who has never spoken. His parents fret endlessly over this handicap, until one day, when he is six years old, he sits down at the breakfast table, puts a slice of bacon in his mouth, and says: "This bacon is cold!" His parents are stupefied! After a minute or so, one of them finally utters: "Why haven't you ever talked before now?," to which the child responds: "Up to now, everything has been OK." As far as the universe is concerned, the quiet evolution of our atomic constituents is far greater. It is as if a child was born, and a million years later he first moves his lips. Nevertheless, in spite of this apparent inactivity, the seeds of our present existence would have been secretly growing.

To learn about this period of relentless cooling, we now turn to the coldest place on earth. Accessible only six months of the year by a four-hour plane flight from the McMurdo Research Station on the coast of Antarctica facing New Zealand, the Amundsen-Scott South Pole Research Station, administered by the U.S. National Science Foundation, is home to the Center for Astrophysical Research in Antarctica. I was surprised when I first learned that while we have been able to send manned missions into space, to the moon and back, no flights, not even military flights, can land at this research station between March and October. The scientists, technicians, bureaucrats, and graduate students living there dig in for the winter. If they need sophisticated emergency health care they are generally out of luck. In 1999, for the first time, an emergency air drop was made to provide medical supplies to a scientist there who was self-diagnosed with breast cancer.

The South Pole in the winter is perhaps the most uninhabitable place on the surface of the Earth. It is dark 24 hours a day, the air is bone dry, and temperatures drop to 100 degrees below zero Fahrenheit, with an average temperature of 76 degrees below zero. For precisely these reasons, it is an ideal place for scientists to work . . . well, at least some scientists. When an astronomer hears the words *dark* and *dry*, he or she begins to salivate.

Three different teams have ventured down to this inhospitable place to make observations as far back as we have ever directly seen. The signal they detect is not visible to the naked eye, or even to sophisticated optical telescopes. Instead, it requires a radio antenna. The scientists at the South Pole are trying to search for minute patterns of noise buried in the cosmic background radiation. The thin, dry atmosphere at the South Pole makes it one of the best places to search for this signal short of going directly into space, which we have done once before and are going to do again in the first decade of this millennium. The noise they are ex-

ploring comes from the motion of electrons interacting with radiation in
regions in the sky that were perhaps 1 part in 10,000 hotter or colder than
the average temperature almost 10 billion years ago. These regions would
contain, we believe, primordial density seeds that would have invisibly
germinated for millions of years, waiting to bloom into galaxies. One of
these seeds, more than 1 million light-years across, would eventually have
enclosed our 16 subatomic particles to make their future interesting.

In 1989, our understanding of these seeds changed forever when
NASA sent up a small satellite to circle the Earth from north to south, escaping the protective cover of oxygen, nitrogen, and water vapor in our
atmosphere that shields us from much of the flotsam and jetsam of
space. The very oxygen atom whose history we are following may have,
at some time or other, helped shield us from the signals that could help
us unravel its history.

The Cosmic Background Explorer (COBE) satellite probably received as much publicity as any nonmanned mission in history, at least
around the time it presented its first results. By measuring the temperature of the radiation bombarding the Earth from space, and comparing
this temperature at different points in the sky to an accuracy of a few
parts in a million, COBE looked back in time, perhaps as far back as we
will ever see. The small hot and cold spots in the microwave radiation
background discovered by COBE emanated from regions that would be
tens of millions of light-years across today, far larger than any galaxy. To
probe for smaller-scale hot and cold spots — whose size makes them
candidates to be the direct descendants of the galaxies we see today —
would require an ability to resolve angular scales smaller than 1 degree
across the sky. The angular resolution of the COBE detector was limited
to about 7 degrees — slightly larger than the size of a major league
pitcher as seen from a batter in the batter's box 60 feet away. Thus
COBE was not sensitive to scales small enough to be associated with the
precursors of galaxies and clusters of galaxies, which is why researchers
have traveled to the South Pole to build telescopes to try. Nevertheless,
the almost imperceptible diffuse lumps COBE uncovered are of fundamental interest in their own right. No physical process during the conventional Big Bang expansion could have created such lumps, since they
are far bigger across than the distance light could have traveled from the

beginning of time to the moment the signal COBE detected was created, when the universe was 300,000 years old. Instead, these primordial patterns would have had to have been imprinted by some process at the very beginning of time. It is for this reason that the American astrophysicist George Smoot said upon first looking at them that it was like seeing "the face of God." To me, however, they just look like smudges.

The remarkable thing about these primordial lumps, these slight density enhancements, is that it is impossible to know that you're inside one of them until a light ray has had time to cross it and provide information that the rest of space outside is slightly less dense. Thus, shortly after their creation, and for almost 1 million years to follow, our 16 particles would have had no way to know what they were in for.

In the early part of the nineteenth century, a German astronomer, H. W. M. Olbers, raised a vexing paradox: If the universe were truly infinite, why was the night sky dark? At first glance, this may not seem like a paradox at all. At night, the sun is not shining! This is true, of course, but as often happens, to properly understand nature we have to reach beyond ourselves. In this case we must remember that our sun is merely one of many billions of stars in our galaxy, which is merely one of many billions of galaxies in the observable universe.

Now, if the universe is infinite in space and time and the farther we look the more galaxies we will continue to see, we are guaranteed that in every single direction we look, we should eventually see a star in some distant galaxy. Of course, you will reason, distant stars are very dim, so even if they are out there, they should be too dim for us to see. But this intuitive reasoning is again misleading. It is true that these stars are dim, but as we look farther and farther out, we should see many, many more stars. It turns out that the number of stars we should encounter increases precisely in inverse proportion to their dimness, so that their total brightness combines in such a way that everywhere in the sky should be as bright as the sun!

The resolution of this paradox does not require any tricks of logic. Instead, there is simply something incorrect about the initial assumption we made in posing it. If the night sky is observed to be dark, then the universe, or at least the stars within it, must have a finite age. This fact should have been clear a century ago, but it was not generally appreciated until it was discovered in the 1920s that the universe *was* expanding, and thus had to have begun in a Big Bang.

If the universe has a finite age, then because the speed of light is finite we can see out only a finite distance. If we can see out only a finite distance, then we are no longer guaranteed to see a star in every direction. It turns out, in fact, that if I draw a line from the Earth to the limits light has traveled since the Big Bang, this line has only a 1 in 1,000 chance of intersecting a star or galaxy. The odds of finding a needle in a haystack are not much worse.

This simple resolution of Olbers's paradox, however, must be modified slightly, but importantly. Because light travels at a finite speed, as we look farther and farther away we are looking farther and farther back in time. If the universe is 12 billion years old, we should be able to see back 12 billion years in time. But 12 billion years ago the universe was very hot. Shouldn't we then be able to see this hot, pregalactic Big Bang directly if we looked out in any direction far enough?

It turns out that we are forever shielded from direct visual observations of the initial Big Bang because as we look farther and farther out, we are guaranteed to hit a wall. Not a solid wall, but an electromagnetic one. Because the likelihood of encountering stars and galaxies as we look back to the Big Bang is so small, a light ray can propagate through the universe of stars to be seen by us today without being impeded. But eventually, if we look back far enough, we reach a time long before stars existed, indeed when atoms did not yet exist, and the universe was a dense gas of protons, nuclei, and electrons, bombarded by radiation of many different types. The light rays that we would like to use to probe this primordial soup cannot penetrate it. When the bare particles of nuclei are unshielded by their cover of electrons, these charged objects interact strongly with electromagnetic radiation. The likelihood that our light ray will be scattered by a proton or an electron before it could reach us approaches 100 percent, so that the primordial soup is opaque, more like tomato soup than consommé. We cannot see into it, we can see only its surface.

But, you might say, shouldn't that surface be glowing hot, and visible, like any stars we may have otherwise wished to observe? Hasn't Olbers's paradox returned, with a vengeance? Well, it takes 12 billion years for the radiation emitted from this surface to reach our telescopes here on Earth, and during this time the radiation has cooled with the expanding universe. Between then and now, the universe itself increased in size a thousandfold, and the radiation cooled by an equal amount. Instead of glowing white-hot, the radiation now shines in invisible microwaves, of the type that the antennae at the South Pole, or on satellites in space, can detect.

We return to our 16 particle building blocks, cooling with the universe after the first few hectic minutes. As long as these elementary nuclei remained electronless, they were helpless to avoid the constant barrage of radiation. Each of our particles was surrounded by a billion particles of radiation. When the universe was a few hundred million degrees in temperature, this radiation took the form of energetic gamma rays. By the time the universe had cooled to a mere 1 million degrees, the radiation had cooled to be X-rays. Nevertheless, a constant bombardment of X-rays, many of them more energetic than those used to provide images of broken bones in hospitals, continued for at least 100 years. In such an intense environment our particles were helpless. The pressure of this radiation gas was immense. The gravity due to any small lumps in the matter distribution would have been powerless to compete with the intense radiation pressure. No structures — stars, galaxies, or even baseballs — could have formed. The radiation gas would have dissipated any primordial lumps on scales small enough to cause matter to coalesce gravitationally. Even neutral atoms, bound by the electric attraction between protons and electrons, could not have withstood this onslaught. The universe, for all intents and purposes, remained completely featureless. Our particles were jostled around like every other proton or helium nucleus in the universe, and had nothing at all to set them apart.

This endless cooling was like an eternity for our particles. Nothing changed. Before, every fraction of a second witnessed something completely new. We wondered which particles would survive the race against time. Now there was all the time in the world. For almost 300,000 years the visible universe was composed of an apparently featureless gas of elementary particles and light nuclei, completely and totally unlike the worlds we see today. The raw materials were there, but not the architecture.

But this was all about to change; formlessness was about to give way to form. After 300,000 years, the temperature of the universe evolved to close to the boiling point of iron. It now glowed uniformly white-hot. The sky would have resembled precisely the universe that Olbers predicted ours should look like, with every point glowing as bright as the sun. But there were not yet any vantage points from which to observe the sky. All there was was sky!

Then, over a 30,000-year period, brief by present cosmic standards, yet still comparable to the entire history of humanity since *Homo sapiens* replaced the Neanderthals, the temperature cooled a trifle more. Suddenly the face of matter was ready to change.

In a hydrogen atom at rest in its lowest energy state, a single electron is bound to its mother proton by 13.6 electron volts of energy. This means that a voltage difference comparable to that produced by a 13.6-volt battery would have to be applied to the atom to drag the electron away. Put another way, the electron in a hydrogen atom has 13.6 electron volts of energy less than a free, unbound electron would have. It thus literally weighs less than a free electron, using Einstein's relation between mass and energy. The binding energy of an electron in helium, containing a nucleus with 2 protons, is even higher, about 20 electron volts. It is thus even easier to capture an electron in helium, and harder to knock it out.

All of our protons, as well as the 2 neutrons and 2 protons tied together in a helium nucleus, continued to play a cosmic tug of war. Incessantly bombarded by radiation, for every billion or so photons scattered by these objects a single encounter with an electron occured. The positive protons attracted the electrons and could momentarily bind with them to form neutral objects, atoms of hydrogen and helium. The electrons would lose energy as they fell in the proton's electric field, and they would emit that energy in the form of light, producing photons that escaped into the radiation gas. As quickly as a neutral atom was formed, however, another photon would collide with the atom and knock the electron free. With 1 billion photons per electron, the cards were stacked against the latter.

But the expansion of the universe slowly worked its wonders. A photon can knock an electron free from its binds to a proton only if it has at least 13.6 electron volts of energy. As the expansion cooled the radiation, fewer and fewer photons could do the job.

Once again, the statistics of rare events makes all the difference. When the temperature of the universe is about 2,700 degrees Celsius, the average energy per particle in the gas is about 0.6 electron volts. This is about 20 times less than the energy required to knock an electron free. However, at this temperature, very rarely, a photon can carry 20 times the average energy. This is like imagining a random wave on the ocean that is 20 times bigger than the average wave. It is not something you would ever expect to see in your lifetime. In fact, the probability for such an event turns out to be about 1 in 10 million. Such a small probability is the type of thing one normally ignores. If you were told that the probability was 1 in 10 million that when you walked out of the house you would be struck by lightning, this would probably not stop you. In fact, the probability of being struck by lightning in a given year is higher than this, yet you are unlikely to know a single person who has had this misfortune, and so we all ignore remote dangers like this. If we didn't, we would be paralyzed.

However, with 1 billion photons for every electron, 10-million-to-1 odds are not good enough to ensure survival. For every encounter a proton had with an electron in which they bound together, a collision with that rare photon almost immediately knocked them apart. But as the universe cooled just a bit more, below 2,700 degrees Celsius, electrons began to gain the upper hand. It takes about 30,000 years for the temperature to drop from 2,700 to 2,400 degrees Celsius, and during this period radiation finally became impotent, and the binding of matter began to take place with impunity. The makeup of matter in the universe shifted almost completely from being composed of charged protons, electrons, and helium nuclei to being composed, for the first time in history, of neutral atoms. From this time on, through the dawn of man until well past the demise of the sun, atoms would reign supreme. The atomic age had begun.

Once matter became neutral, all the rules of the game immediately began to change. The radiation that so efficiently interacted with protons and helium nuclei now found itself virtually irrelevant. The neutral atoms coexisted with the ever cooling radiation gas, but they were largely impervious to its various charms. The energy stored in this background radiation continued to dilute in proportion to the energy stored in atoms so that it played a progressively less significant role in the cosmic balance. Atoms and the background radiation photons now followed separate destinies. Indeed, it would take, in our own remote corner of the universe, more than 10 billion years before the very existence of this ubiquitous background radiation would be uncovered by conscious lifeforms — even though the background had been with us from the beginning. From the moment almost 12 billion years ago when radiation and matter first decoupled until the present time, the average photon in the cosmic radiation background would never again significantly disturb a single atom of matter in the universe.

As the universe cooled, it also began to get darker. The uniform radiance of the radiation gas receded from blue to red, and beyond. By the time the universe was 3 million years old, the afterglow of the Big Bang had begun to shift away from visible light. By 10 million years into the expansion, less than 1 percent of the radiation bath would have been visible, glowing with an eerie red light, if anyone had been around to see it. Nevertheless, there was still enough energy contained in the cosmic background radiation that the brightness of the night sky then, without a single star to enlighten it, would have been about 1 percent as bright as the daytime sky is today on Earth. This means that the primordial gas was about 1 million times brighter than the night sky is today on a moonless night far away from city lights.

In the glow of the cosmic night 300 millennia after the beginning of time, it might have appeared that our 16 lonely particles, surrounded by their electronic sheaths, were destined to disappear into an ever-growing darkness, ultimately blacker than a storm cellar in a hurricane. But this was not to be the case. By yet another cosmic coincidence, just as the sky became as dark as the night sky is today, the long march toward oblivion was arrested.

Once neutral atoms became the stuff of matter, gravity finally and completely took over the show. The long-range repulsion between the single charged particles disappeared when they bound together into neutral atoms. And if the photons were no longer scattering on the atoms, with each passing year their own intrinsic pressure became yet more irrelevant.

Suddenly, imperceptible clumps of matter could begin to respond to the demands of gravity. Early in the history of the universe, perhaps at the same moments that baryons were themselves created, a network of imperceptibly diffuse regions with a slight excess density was laid out, waiting. Three hundred thousand years later, as the cosmic radiation de-coupled from matter, the gravitational pull within these regions — with density enhancements of less than 1 part in 10,000 compared to the av-erage density of matter and radiation — would have left an imprint on the radiation bath we now detect as the cosmic radiation background. Once again, electrons played a key role in the process. Just as matter was becoming neutral and about to respond to gravity in the regions that had a slight density excess, the electrons that were about to be captured into atoms scattered one last time off of the radiation bath, disturbing it ever so slightly. The result was an effect that is almost imperceptible, unless, say, you are willing to brave the coldest place in the world for six months at a time, and bring with you the most sensitive microwave detectors on the planet.

Aside from this minute effect, these regions with excess density would continue to expand uneventfully along with the universe not merely for another 300,000 years, but for another 100 million years. This was perhaps the longest period in the entire history of the universe until the present time in which nothing of observable significance would hap-pen. For longer than the time it might take for life to spring from the primordial ooze on Earth, for far longer than it would take from the time humans first walked upright to the time they would circle the globe in rocketships, the universe, clumps and all, simply expanded. The clumps, however, would expand at a rate just a smidgen slower than the regions outside them. In this way, every time the universe dou-bled in size, these regions would not quite double in size, and thus the density contrast between them and the outside would continue to ever so slowly grow.

As long as our 16 protons and neutrons were electronless, the intense pressure of radiation stopped them from responding to the growth of the background density enhancements. Once the particles formed neutral atoms the gravitational attraction of the underlying clumps began to make itself felt. The 13 atoms — 12 hydrogen atoms and 1 of helium — that would one day join together to form our oxygen atom, became caught up in an expansion within an expansion, from which there would be no escape. But for 100 million years they rested.

7

THINGS THAT WENT BUMP IN THE NIGHT

We will now discuss in a little more detail the struggle for existence.

CHARLES DARWIN

For the newly born atoms in the emerging darkness following the Big Bang, the struggle between pressure and gravity was about to begin in earnest. It would continue for all eternity, governing the ultimate destiny of every object in the universe. The outcome was never in doubt. Gravity will eventually win.

Darkness hides many desperate struggles. Buried somewhere in our ancient id lies a primal fear of night as old as humanity itself. Awe-inspiring landscapes during the day become menacingly eerie in the dark. At night, every boulder, every tree, every gully, every alley provides cover for a potential nasty surprise. The progress of human civilization follows, in large part, the effort to conquer night. Wood fires and oil lamps, succeeded eons later by kerosene and gaslights, and then electric lights, helped hold the night at bay, removing it beyond arm's length, to a safe distance. Things that go bump in the light are curiosities. Things that go bump in the night keep us under the covers.

In the great cities of the world, we have detached ourselves from night. If you are a city dweller who doesn't believe this, travel at least a hundred miles into the countryside, mount the highest hill, and stare at the sky. It is not the same sky at all. In a city, the stars overhead glitter like lights on a distant rooftop, and the sky begins beyond the horizon. On a clear night in the mountains, you become a *part* of the sky. The stars reach out and touch you, and suddenly, you feel the embrace of a galaxy.

One can only imagine the feeling the first conscious humans had when they looked up at night. Did they draw comfort from the giant arc of the Milky Way overhead? The writings of the world's early religions suggest the sky provided evidence that there was order in the world, that all things had a place, and that we were being watched over by intermittently vengeful or benign deities. But most important, the separation between Sky and Earth had not yet clearly taken place.

The Egyptians had a mythology full of heavenly beings, but for them the stars themselves also had a direct connection to the land. The central event in Egyptian life was the annual flooding of the Nile River. This occurrence determined the entire life cycle of the community, with flooding, subsidence, and the planting of crops dividing the year into three seasons. The singular event that heralded the coming of the floods was the *heliacal rising* of the brightest star in the sky, Sirius. This star remains close to the sun for part of the year, and is hidden in its glare. But after a lengthy absence, it reappears one morning in the dawn sky. This annual rising reoccurred just as the Nile began to flood. The two events were connected not merely in the establishment of the Egyptian religious calendar, but in the sense that events on Earth were intimately connected with the stars.

Even today, the Mursi peoples of southwest Ethiopia associate the annual flooding of the Omo River with the heliacal setting of stars in the famous Southern Cross, just as they connect the flooding to the flowering of various plants. The Misminay villagers in the Andes take this connection to its logical extreme. The Milky Way is viewed as merely a celestial extension of the Vilcanota River, circulating waters from the heavens to the Earth.

While we now realize that there is no direct connection between daily events on Earth and the motions of stars in the sky, the Milky Way *is* a sort of cosmic river, whose undulations have been responsible for our own existence, just as they may one day govern our demise.

After all, our galaxy is not static and unchanging, even if it may appear to be on the timescale of human civilizations. The stars that light up the sky are to the Earth like ships passing in the cosmic night. If we consider one rotation of the Milky Way galaxy, which takes about 200 million years, as a single galactic "day," these stars are our neighbors for only a few hours. With velocities of up to 200 kilometers per second relative to us, after one galactic revolution some stars will have moved 20,000 light-years distant, well out of the range of the night sky visible to the naked eye. Since the formation of our galaxy there has been time for perhaps 50 full revolutions of the Milky Way, more than enough time for many stars in the galaxy to traverse the full distance across the galaxy from any other star. And before the spiral disk of the Milky Way was sculpted, a cosmic dance of "waves" of gas repeatedly crisscrossed the entire region that would eventually become our galaxy.

The 16 particles that make up the nucleus of our oxygen atom today were thus not in any sense neighbors at the remote beginnings of eternity, or for much of the time in between. They literally may have been galaxies apart.

The complex future yet to take place could then not have even been imagined for our primordial hydrogen and helium atoms. For them, the future looked secure, and simple. With each passing century, for a million centuries, the bombardment of photons relaxed ever so slightly, neighboring atoms moved slightly farther away, and the descent into darkness would have appeared to march relentlessly forward. The 12 hydrogen atoms and 1 helium atom destined to eventually join together were spread out over a distance of more than 1,000 light-years, about 1 percent of the present size of the disk of the Milky Way galaxy. That these objects should eventually coexist within a region far less than a trillionth of a meter across would have appeared unlikely at best then. This is the wonder of an old and ever-expanding universe: Unlikely events are nevertheless bound to occur.

Eight of our hydrogen atoms and 1 helium atom found themselves located in a clump 500 light-years across. This clump, while dominating the local mass, nevertheless had at that time a density of only slightly greater than the average mass density of the universe. It would have been very difficult to know that this region was in any way different. The effect of this slight density excess was, as I have described, merely to slow by an imperceptible amount the cosmic expansion of material in the region.

Imperceptible at any single time, that is. Over the course of 100 million years, even a small effect could build up.

From the beginning of this period, at 300,000 years, over the next 100 million years the universe would expand by a factor of about 50. The region containing the 9 atoms in question, however, would grow by merely a factor of 40, growing to span 20,000 light-years across, comparable to the present distance between the sun and the center of the Milky Way. This slight difference in growth factors would mean that the matter density within this cloud was now twice as great as that of the universal average value.

Self-determination now became possible. The gravitational attraction of this huge diffuse cloud of matter was now sufficiently great that it could break away from the background cosmic expansion. But putting the brakes on perhaps 40 billion solar masses of material is not so easy. The region would continue to grow in size by a factor of 2 over the course of the next 200 million years before its expansion would cease, not just momentarily but for all time. One of the first among the burgeoning set of "island universes" — and one of the seeds of our Milky Way — now began to take form.

For our particles, the fact that their chains had grown to bind them would still not be obvious. Indeed, the term *gravitational collapse* is, within an expanding universe, often a misnomer. The effect of gravity serves primarily to halt the expansion. Whether collapse ultimately occurs, and in what form, depends on a host of other factors. By the time the pregalactic sphere of gas stopped expanding, its density, while about 6 times that of the background region, was only about 20 times less than the ultimate mean density of the Milky Way galaxy today. As a whole, the region might compress by at most a further factor of 2 to 3 in size over the next 10 billion years.

That the process of gravitational collapse is not merely the simple time reversal of the previous expansion is fortunate. If it were, then atoms would reheat to ions, which would then reheat further, breaking up nuclei into protons and neutrons, then quarks, and so on. The complexity of matter in the universe would never move beyond single atoms of hydrogen and helium.

Why, you may ask, is gravity so inefficient? After all, if I drop an egg out of an airplane, gravity does a pretty good job of bringing it all the way to the ground and keeping it there. But what if I drop a Superball from an airplane? Better still, what if I drill a hole right through the Earth and drop the egg? In this case, the egg will fall to the center of the Earth, but it will not stop there. As it falls, it will build up speed, so that it will rocket past the Earth's center and come out the other side, rising to a height very nearly equal to the height it was dropped from on this side of the Earth.

So too for the atoms in the collapsing cloud 40,000 light-years across. If the density in this cloud were completely uniform, and there were no random motions of the individual gas particles, one could show that all of the particles in the gas would stop their outward flow at the same time, and begin heading toward the center together. The outer particles would travel faster than the inner particles, so that they would all reach the center at the same time, in a massive collision. However, this situation is the exception, rather than the norm. Most clumps of gas are not uniform in density. If they were, then at the edge of the clump there would be a sharp discontinuity between the density of the clump and the density of the background. Solid objects may exhibit this behavior, but not diffuse pregalactic gas clouds.

For the cloud in question, all the atoms in the cloud would not collapse together. Inner shells of material would begin to collapse before outer shells. Moreover, the individual particles in the gas would begin this collapse inward with additional random peculiar motions characteristic of the small, but nonzero, temperature of the gas. As gravity pulled them inward, they would thus not head directly toward the center of the configuration, but they might aim slightly askew. As a comet does when it heads toward the sun, such particles would miss the center, and instead move in highly elongated orbits.

Also, remember that when the gas cloud first stopped expanding, its average density would have been only about 1 atom per 10 cubic centimeters. The distance between atoms was thus over 100 million times their individual size! The net effect of this fact, combined with the slightly noncentral trajectories of the individual particles, was that each collapsing shell of material could pass through the central regions of the spheroid and come out the other side — like our egg through the Earth.

As the particles "fall" toward the center, they speed up, and then slow down on their way out.

When I was a kid, we used to play a game called Red Rover. Two groups of children would face each other, with one group linking hands. This group would call out the name of one of the kids from the other group, who would then have to run fast and attempt to break through the human chain. Those who did continued again, while those who didn't joined the ever-enlarging chain.

The process by which stars formed out of the emerging protogalactic gas clumps is analagous. First, random motions of individual atoms might cause their trajectories to converge. While the particles would emerge from these encounters as quickly as they had entered, the effect of innumerable such interactions would be to redistribute gravitational energy among them. Such clumps would become hotter — the atoms within them would be moving more quickly, and they would, as a result, not collapse further. Every now and then, however, much more rarely perhaps than schoolchildren playing Red Rover, 2 atoms would have a head-on collision.

I have been discussing this complex of atoms as a gas, but you may wonder whether this is reasonable. After all, the density of particles here is far smaller than in even the greatest vacuum that can be created on Earth. The density may be 1 particle per cubic centimeter in this cosmic gas, whereas in a vacuum tube on Earth the mean density is usually about 1,000 billion times greater. The difference here, as perhaps elsewhere, is that size matters. The mean distance between collisions of an atom in the cosmic gas cloud is immense, greater than the size of our solar system. The size of the cloud can be much larger than this, however, so that the atoms "regularly" collide as they pass through the cloud. When atoms can collide and redistribute their energy a number of times before they could otherwise make a single traversal of the system, they behave as a gas, with pressure, shock waves, and the like.

Whenever 2 atoms in the gas collide, something noticeable — literally — may happen. The electrons orbiting the atoms can become excited, changing their orbital configurations. Shortly thereafter, they will relax back to their ground states, emitting light in the process. Alternatively, if they are moving slowly enough and come close enough together, the individual electron clouds surrounding each atom might join together, releasing energy in the process, and the hydrogen atoms can bind in pairs to form *diatomic* molecules, at this point the largest microscopic structures to appear in the universe. The net effect of either process would be for the atoms, and now molecules, to be able to slowly convert their energy of motion into light.

It is vitally important for their future that our atoms of matter can so dissipate their energy. Without this possibility, gravity would be impotent. It could halt the local expansion of gas clouds, but it would never produce the rich structure we observe throughout the universe today. Indeed, we would not be here today to observe anything at all.

This primeval gas, made almost solely of hydrogen and helium, is different from the gas that makes up interstellar space today. On the surface, the primeval gas might resemble what we see now, but there are subtle but important differences. In particular, there are essentially no trace amounts of any elements heavier than helium in the former. The many different electrons in heavy atoms can be excited in many different ways. These in turn provide efficient "refrigerators" that can convert atoms' stored energy of motion of atoms into radiation, keeping the whole gas cool as it collapses.

In our primordial gas, however, there was just hydrogen and helium. Helium, being a so-called noble gas, is inert. It does not bind to form molecules. Hydrogen, however, can bind together in pairs, as I have described, to form molecules. Molecules are, in a sense, infinitely more complex than atoms. This is because the particles in a molecule can perform complicated dances around each other. The energy to excite such dances is much, much smaller than the energy required to excite individual electrons in each atom to change their motion.

If hydrogen could not bind together into molecules, the process of gravitational collapse of our primordial gas would have been resoundingly different. Because the hydrogen molecules can be excited so easily, they can efficiently cool the gas. The energy that would otherwise be converted into motion of the atoms as they stream through the center of dense regions is instead converted into rotational and vibrational energy of the molecules, which is then converted to radiation as the molecules relax, emitting photons of infrared light. These photons can escape out into space, leaving in their wake a cooler system. In this way, the denser molecular clouds in which collisions occur most frequently can most efficiently convert gravitational energy into radiation, losing energy and slowly increasing their density, without heating up in the process.

There is another fly in the ointment, however, and one whose origin we do not fully understand at the present time. As we observe light emitted by objects in our galaxy, and other galaxies, it has become clear that our galaxy is threaded by a large, if weak, magnetic field. Magnetic fields are generated by the motion of electric charges. Our Earth is a magnet, for example, because of electric currents flowing in its molten iron core.

Although we do not yet understand the origin of the coherent magnetic fields on galaxy-size scales, we do understand that these can play a key role in governing how primordial clouds can collapse. Charged particles not only create magnetic fields as they move, their motion is itself affected by the presence of background magnetic fields. In such a field, charged particles are not free simply to respond to the stresses induced by gravity. The particles, it turns out, can move much more easily *along* the direction of the magnetic field than they can perpendicular to it.

By this point, almost all the atoms in the collapsing clouds are neutral, but some very small fraction of atoms remains ionized. These charged particles pin down the magnetic field lines, and in turn have their motions constrained. Thus any collapsing cloud has to contend with two impediments: the outward pressures caused by heating of matter as it collapses, and the inhibiting role of primordial magnetic fields. The two factors together must determine why and how matter in the universe collapses into stars and galaxies.

In this emerging struggle between pressure and gravity, size is *everything!* If you cut up the Earth and separated it into individual 1-gram pieces, each piece would have a minuscule gravitational attraction to each other piece. The speed of the individual atoms at room temperature would be over a million times the speed necessary to escape the gravitational pull from any of the individual pieces. On the other hand, a system that had the same average density of the Earth today but that had a mass equal to our entire galaxy, would form a black hole, out of the gravitational pull of which not even light could escape. The old saying is really true. The bigger they come, the harder they fall.

This behavior reflects a simple scaling relation in physics. The pressure of a gas, which is what holds it up against collapse, is proportional to the temperature of the gas and its density. If the gas is kept at a constant temperature and density, the overall thermal energy stored in the gas, producing the pressure that resists collapse, is thus simply proportional to its volume, which grows as the cube of its radius. For the same system with constant density, however, the gravitational energy, related to the gravitational attraction that induces collapse, grows as the fifth power of the radius of the system. If one continues to increase the radius of such a system (and hence its overall mass), eventually the gravitational energy must beat out the thermal energy. Once the gravitational energy wins out, the system will start to collapse.

One can turn this argument around. A larger system, containing more mass, will thus be more likely to collapse than will a smaller, less massive system having the same density. For example, a primordial gas cloud with mass equal to that of the sun spanning a size of about 1 light-year across, with a temperature of, say, 10 degrees above absolute zero, will not collapse under its own gravity. A cloud with the same density, however, and a radius 10 times larger, comprising about 1,000 solar masses, will contract.

If this is the case, how come the sky is not full of stars with masses 1,000 times the mass of the sun? Well, let us consider the 1,000-solar-mass cloud again. As it contracts, its average density increases. If the molecular hydrogen refrigerator continues to operate efficiently, the temperature of the cloud will not increase as it collapses. At a certain point, if it collapses by a factor of 50 in size, then the density will have increased by a factor of 125,000! A solar mass of material will now be contained in a region about 1/50 of a light-year across. This region now satisfies the criteria for gravitational collapse and can separate out from the background and collapse on its own.

In this way, as larger clumps begin to collapse due to gravity, smaller subclumps begin to be able to condense out of the larger system and collapse. The evidence for this is that stars tend to be born in large associated groups rather than as isolated systems.

What stops this process of fragmentation into smaller and smaller subclumps? Eventually, the density required for the smaller clumps to condense is so great that the system no longer remains transparent on this scale to the radiation emitted by the atoms and molecules as they collide. Once the light cannot freely escape, the temperature of the system begins to rise, its internal pressure begins to increase, and the process of collapse is, at least temporarily, halted. In the process, a protostar is born.

There are two other factors that affect the degree to which gas clouds can continue to collapse to eventually form stellar systems. The first is the role of ubiquitous galactic magnetic fields that tend to restrict the movement of atoms once they become ionized. The second is the fact that the separate regions in these clouds can be moving with respect to one another, with a small net rotation. As the gas clouds begin to collapse, like a rotating figure skater pulling her arms in, the system as a whole will begin to rotate faster and faster. If massive gas clouds start off with rotation velocities as small as 1 kilometer per second, about 1/200 the rotation speed of the sun around the Milky Way at the present time, then well before clouds of the densities described here have collapsed sufficiently to form stars, they would be rotating at the speed of light! In order to avoid this unphysical situation and to allow further collapse, clouds must thus shed some of their rotational velocity. They can do this by several techniques: by colliding with nearby clouds, by pumping out their rotational energy to the surrounding magnetic fields, and most important, perhaps, by shedding the fastest-moving particles in a stellar wind as they collapse. Thus, perhaps 30 to 50 percent of the total mass of a 30-solar-mass cloud may be spewed out in such a stellar wind as the remainder continues to collapse. This outflow of material will smash against the material surrounding the cloud, inhibiting collapse in the region of the emerging protostar. It is as if the emerging star buries itself deep inside a cocoon of its own making. For less massive clouds, the faster-rotating material shed by the collapsing cloud remains in orbit around it, leaving the fodder for possible smaller objects to form.

The 8 hydrogen atoms and 1 helium atom in one of the huge pregalactic gas clouds have now found themselves in quite different circumstances. Four of the hydrogen atoms and 1 helium atom are moving through a molecular cloud containing 1,000 solar masses, and 300 light-years across. Remember that these 5 atoms, hurtling through the unrelenting darkness, are now a part of an oxygen atom, which may be in the air you are now breathing. Part of you was there, spread out over a larger distance than the sun has traveled around the galaxy since the dawn of humanity.

Contracting over the course of about 10 million years, this system decreases in size by about a factor of 10, so that it is now 30 light-years across. By now the 4 hydrogen atoms have become bound with other hydrogen atoms into 4 molecules. At the time they decoupled from the expansion of the universe, perhaps 100 million years earlier, the background temperature of the radiation was about 50 to 60 degrees above absolute zero. While some momentary heating of the gas took place as collapse began, once the molecular refrigerators began to operate the cloud cooled to a temperature of perhaps 10 degrees above absolute zero — over 440 degrees below zero Fahrenheit. Under such conditions this cool molecular cloud continues to collapse.

Our hydrogen atoms, now molecules, collide more frequently as the gas density increases. There are now about 10,000 atoms per cubic centimeter, a density enhancement of over 100,000 since the expansion of the proto-galaxy stopped. The rate of collisions between particles is more than 1 billion times larger than it was at that point, but conditions for our atoms are still relatively benign. As molecules, they are tumbling, jiggling, and rotating through a diffuse gas, their shared electrons dancing in unison around them, emitting radiation in response to each external shock, so that they remain cool. As far as they are concerned, the peace of the past 300 million years largely persists.

Within this emerging clump, our atoms approach each other at a distance of about 1/5 of a light-year, over 100 times the size of our present solar system. While it would take them more than 1 billion years to traverse the gap separating them via random thermal motion, they are in a collapsing cloud and thus are already destined to be on a collision course. At a density of 10,000 atoms per cubic centimeter, a clump of gas 30 times the mass of our sun has now begun to separate out from the larger cloud and collapse inward.

Our hydrogen molecules and helium atom are now in gravitational free fall. Accelerating inward, they traverse over 50 percent of the distance separating them in perhaps 1 million years. Collisions with their neighbors, followed by emission of radiation, continue to keep the hydrogen molecules cool, and the helium atom is carried along for the ride, its thermal energy held in check by the cooler surrounding hydrogen molecules.

Things now begin to heat up for our particles, however, both literally and metaphorically. As the density of the system continues to increase, more and more of the radiation emitted during collisions is reabsorbed by the molecules. They slowly get hotter in response.

Nevertheless, on and on the free fall continues. Imagine continuing to fall for more than 100,000 years! During this time, in order to keep from heating up, the gas must radiate tremendous amounts of energy. Finally, when the cloud reaches a size about half that of our present solar system, after having compressed by slightly less than a factor of 1,000, the free fall stops. The gas is so dense (about 1 ten-billionth as dense as water) that the infrared radiation emitted by the infalling molecules gets trapped and reabsorbed before it can escape. The associated heat energy now trapped in the gas causes it to exert a pressure resisting further collapse.

During the many years of collapse, the cloud has been emitting radiation in order to remain cool. The amount of radiation emitted is equal to about half the gravitational energy lost by the infalling mass, and this is tremendous. The total energy radiated during this period is comparable to the total energy radiated by our own sun over the course of the past 1 million years.

Since this collapse takes place over a period somewhat longer than 100,000 years, this cloud is therefore, on average, more luminous than our own sun! There are two big differences, however. Our sun shines in visible, ultraviolet, and X-ray radiation. The collapsing cloud is emitting primarily in the infrared. More important is the fact that our sun has been shining with a roughly constant luminosity over the past 1 million years or so. But the radiation emitted by the cloud increases as its temperature increases. Almost all of the radiation is therefore emitted during the final stages of collapse. During a short period, lasting perhaps a decade, our cloud shines with a brightness in excess of 10,000 suns!

It is during this final phase of collapse that our atoms begin to respond noticeably to the fact that their universe is no longer cooling. The heat produced in the cloud builds up until there is sufficient energy to break apart the hydrogen molecules, requiring a temperature of a few thousand degrees. The atoms are back to where they were 300 million years earlier, shortly after they captured their first electrons to become neutral. The difference is that in the early universe, radiation was everywhere, and far more dense than matter. Now matter is king. In the collapsing cloud, the density of matter is 1 billion times greater than it was when the universal radiation bath had a temperature of 3,000 degrees.

The moment our gas cloud fully traps its own radiation, its luminosity drops abruptly, and its collapse begins to slow down dramatically. Everything changes. The pressure of matter now begins to counter the pull of gravity. The normal gravitational free-fall-collapse time of a gas cloud of this size is several years. Instead, this dense cloud can now survive for at least 10 million years without succumbing to gravity. Our atoms are about to become part of a star.

PART
TWO
VOYAGE

I am standing on the threshold about to enter a room. It is a complicated business. In the first place I must shove against an atmosphere pressing with a force of fourteen pounds on every square inch of my body. I must make sure of landing on a plank travelling at twenty miles a second round the sun — a fraction of a second too early or too late, the plank would be miles away. I must do this whilst hanging from a round planet, heading outward into space.

SIR ARTHUR STANLEY EDDINGTON

8.
FIRST
LIGHT

*I am aware that many critics consider
the conditions in the stars not
sufficiently extreme . . . the stars are
not hot enough. The critics lay
themselves open to an obvious retort:
we tell them to go and find a hotter
place.*

SIR ARTHUR STANLEY EDDINGTON

In 1854, the British physicist William Thomson, later known to the world as Lord Kelvin, discovered that the sun was too old to shine.

Thomson and the distinguished German physicist Hermann von Helmholtz independently concluded, using the known laws of physics at the time, that the sun could not have been burning as brightly as it does for more than between 20 and 100 million years. Needless to say, this result was embarrassing, because the Earth was already known, from the geological record, to be at least an order of magnitude older than this, and perhaps 100 times as old.

Nevertheless, embarrassing or not, these two gentlemen had at the very least extended the explicable lifetime of the sun by a factor of 10,000. The natural *gravitational lifetime* of the sun is on the order of 48 minutes — this is the time it would take the material in the sun to collapse to the center if gravity were not offset by the braking pressure of the hot gas. But something had to act to keep the gas hot. The German physician J. R. von Mayer had earlier calculated that even supposing the sun's core were made of coal and supplied with enough oxygen to fully burn it, it could burn with the observed brightness for only 1,000 years or so.

Thus the Kelvin–Helmholtz calculation, as it has become known, represented a milestone. It is actually a rather simple estimate, the kind that can be done on the back of an envelope. Simply take the total energy that can be turned into internal heat and pressure by the gravitational contraction of the sun's mass — which is about half the total gravitational energy released, the other half being radiated into space — and divide this by the rate at which the sun produces energy. The result is a timescale of 40 million years.

Perhaps the most significant aspect of this result, even if it yields a lifetime that is far too short, was that it implied the sun could be contracting before our eyes, but so slowly that in the course of a single human lifetime, or in the course of an entire human civilization, the amount would be unnoticeable. The sun, and by inference the stars, need not be immutable.

But something was still clearly missing. At the present time in the United States, with the resurgence of "creationist" fervor, such a conundrum might have led to an outcry for a revision in the public school curriculum. For if science couldn't yet explain why the sun was still shining, then perhaps a 6,000-year-old Earth might yet be viable! But this was the late nineteenth century. Science was in its heyday, powering the Industrial Revolution, and there was faith, not that divine intervention would be required, but rather that science could eventually uncover the secret process fueling the stars.

This faith required almost 100 years for its vindication. In 1939 the physicist Hans Bethe demonstrated that Sir Arthur Stanley Eddington's belief in the efficacy of nuclear furnaces, expressed in the quotation at the beginning of this chapter, was not misplaced. Bethe showed that a series of nuclear reactions, starting with 4 hydrogen nuclei, could produce a helium nucleus in a process that would release 10 million times the energy that an equivalent quantity of coal does when it burns. With 10 million times the available fuel energy, therefore, the sun could survive 10 million times longer than Mayer's earlier 1,000-year estimate, or about

10 billion years before succumbing to the inevitable pull of gravity. Kelvin and Helmoltz's upper limit was wrong by a factor of 1,000 or so because they had no knowledge of the existence of the atomic nucleus, and the incredible energy stored therein.

Bethe had firsthand knowledge of the awesome power trapped inside atomic nuclei. As head of the theory division in the Manhattan Project, he played a key role in the wartime development of the atomic bomb at the Los Alamos laboratory. As terrifying as the first nuclear explosion was, the process that generated the energy behind it, nuclear *fission*, generates far less energy per unit mass than the *fusion* process that powers the sun. It was only after the war, with the development of the hydrogen bomb and the explosion called Mike, that we first harnessed the energy of the sun, albeit in an uncontrolled way. We have yet to generate controlled fusion reactions in the laboratory that produce more energy than the energy required to generate them. If we do, these may replace the sun as the prime fusion energy source for life on Earth.

You might wonder why I am bothering with fusion again here. After all, some of our atoms went through this very process to form helium in the first minutes of the Big Bang, didn't they? Yes and no. Remember that in the early universe, neutrons had not yet decayed, and could combine with protons to form the building blocks for the eventual production of helium. After the universe was more than about 10 minutes old, however, all free neutrons would have long since decayed away. Outside of the neutrons contained in helium (1 or 2 depending on the isotope), there were essentially only protons left over, in the form of hydrogen, with which to build stars.

Let's put it another way. In the early universe, creation of helium was a race against time. The universe had just 10 minutes in which to convert protons and neutrons to helium before losing free neutrons forever. The fact that this could actually happen in such a short time was a minor triumph. Now, almost half a billion years later, the universe had all the time in the world to further the process. However, it was now, on average, 1 billion times colder, and moreover, half the building blocks of atoms no longer roamed freely in space.

How can helium be built out of only protons? Although collisions between protons and neutrons that lead to the two particles' binding together are not difficult to imagine, protons carry electric charge, so that 2 protons that approach each other should experience a repulsion, not an attraction. How can nature overcome this natural disadvantage?

It was against the backdrop of these sorts of questions, although they were much more vague then, that Eddington framed his famous retort in 1926. Some skeptics at that time doubted it would be possible for nature to generate conditions inside of stars sufficiently exotic for any new subatomic physics to operate. Of course, this skepticism was not based on any clear understanding of nuclear physics. In fact, with hindsight, and with such an understanding, one's initial skepticism may be increased, not decreased!

Remember that primordial nucleosynthesis occurred over the course of about 1 minute, when the net temperature of the universe exceeded 1 billion degrees! The surface of our sun is only about 6,000 degrees Celsius, and no one then, or now, had the temerity to suggest that the conditions inside the sun could increase the temperature there by a factor of 1 million compared to the value on its surface.

In fact, as is often the case in science, what appears to be a fundamental problem is later seen to be a remarkable blessing, once one properly understands the issues at hand. For if the temperature inside the sun were not 1,000 times smaller than the temperatures available during Big Bang nucleosynthesis, and if neutrons were freely abundant, stars that condensed out of collapsing gas clouds would have been able to convert all their hydrogen to helium in the course of minutes, not millions or billions of years. The lifetime of a star would be comparable to that of a fruit fly, and life in the universe, even as primitive as that of fruit flies, would never have been a remote possibility.

We now return to a time well before stellar nuclear reactions began, in our collapsing cloud, when our atoms are in the final stages of gravitational free fall just as they begin to form a protostar. The temperature is increasing, and one by one, hydrogen molecules (H_2) are dissociating back into atoms. Things are speeding up as every year passes. The early, uneventful eons of collapse culminate in a decade or so of frenetic activity. Atoms are rushing together, and the temperature increases. The intense infrared radiation field begins to heat up matter on the outskirts of the flow, blowing it back into space. Thirty solar masses of material that began the inward collapse may be only 15 solar masses of material by this point, and perhaps only 10 solar masses of material will complete the journey inward. In the final decade of free-fall collapse, the ball of gas will emit more energy than our sun has emitted since human civilization began on Earth up until the present time.

It is a turbulent era, but this is nothing compared to what is to come. The temperature of the collapsing gas ball continues to increase with every year, more of the molecules become dissociated, and more of the radiated energy gets trapped in the ball, heating it further. Finally, now the size of the inner part of our solar system up to Mars, the collapsing ball slows down, as the ball becomes fully opaque to the emitted radiation and the trapped energy increases the pressure further.

By this point, fully half the energy released in each moment of collapse is turned into energy of motion of the atoms, and half is radiated outward as infrared radiation of ever increasing frequency as the protostar continues to heat up further. Yet our protostar continues to collapse. The temperature and pressure of the turbulent matter still cannot rise fast enough for the pressure to counteract gravity's relentless attraction.

The reason for this inability of pressure to combat gravity is the fact that our atoms (remember that all the molecules have by now been dissociated into atoms again) are largely neutral objects. Recall that it takes a temperature of between 10^3 to 10^4 degrees for radiation to be energetic enough to ionize hydrogen. By the same token, as long as radiation energy can go into ionizing atoms, it is impossible for the temperature of the matter to rise above 10^4 degrees. It is like heating up ice. Ice melts at a temperature of 32 degrees Fahrenheit. If you continue to add heat to the ice, this heat will go into the melting process, and the temperature will not rise above 32 degrees until the melting process has been completed. So too with a protostar. Until the hydrogen and helium making up the collapsing cloud have become ionized, the temperature cannot exceed 10^4 degrees. Until the temperature can rise again in full response to the input of energy from the collapse, the pressure cannot build up to counteract it, and the collapse continues, albeit more slowly than during the free-fall stage.

As the protostar continues to collapse in this way, the luminosity it emits will go down as its surface area decreases. For a protostar of 15 solar masses, the luminosity will decrease by an order of magnitude over a period of less than a year. This luminosity will still be immense, over 10,000 times the luminosity of the sun, powered completely by gravitational contraction, with no other internal source of energy.

Finally, with all atoms ionized, the temperature and pressure are free to rise in response to collapse. The collapse slows dramatically, so that it is now more appropriately called a contraction. The ball of hot gas is now on the order of the size of the Earth's orbit around the sun. For the first time since the earliest moments of the Big Bang, the gas is in full *hydrostatic equilibrium*, with pressure balancing gravity. Of course, because there is not yet any internal energy source, the protostar, which is radiating away energy, must continue to contract in order to maintain its pressure. Thus the system is not yet truly "static" in any sense. Nevertheless, the timescale for significant contraction for our massive cloud is now on the order of a century, rather than a year. In any case, once the contraction slows under hydrostatic equilibrium, it is reasonable to begin to call our object a star.

During this contraction phase, the nascent star encounters a huge problem. In 1961 the Japanese astronomer Chushiro Hayashi pointed out that in order to maintain the pressure necessary to momentarily counter gravity, a contracting star would have to maintain a high temperature so that its surface temperature does not become too cool. But these objects have a huge size, and with a high surface temperature, on the order of a few thousand degrees, their luminosity must be immense. Because the inside of the star is at this point opaque to radiation, the only way to transport energy to the surface fast enough is for huge masses of gas to flow up to the surface, by convection. Our star becomes similar to a huge pot of boiling gaseous oatmeal, with mammoth turbulent eddies and bubbles.

As this huge convection stream is set up, heating the outer part of the star, the star's luminosity jumps dramatically for a very short time, achieving a value in excess of 100,000 times that of the sun for a period of less than a year. This radiation burst is still peaked in the infrared, but now the near infared, much closer to the visible range. As the surface heats up, the emitted radiation moves into the visible band. The star has, from a human perspective, begun to truly "shine."

This intense glowing ball of gas is surrounded by a huge cocoon of gas expelled during its collapse, heated in turn by the radiation emitted from the star's surface. Our 4 hydrogen atoms and 1 helium atom have survived the infall, and now find themselves located near the surface of the star. But not for long. The convection flows drag them deep inside, heating them further, then pushing them out again, as the massive star contracts in size. In this way, they experience the totality of the cosmic ride afforded by the evolving star. Our atoms get alternately heated to hundreds of thousands of degrees as they fall, and cool to perhaps 3,500 degrees Celsius as they approach the surface.

Perhaps no better description exists of the inside of a nascent star than that given by Sir Arthur Stanley Eddington, as he struggled to imagine by what process the ultimate star might be powered. In 1926 he wrote:

> The inside of a star is a hurly-burly of atoms, electrons, and aether waves. We have to call to aid the most recent discoveries of atomic physics to follow the intricacies of the dance. We started to explore the inside of a star; we soon find ourselves exploring the inside of an atom. Try to picture the tumult! Disheveled atoms tear along at 50 miles a second with only a few tatters left of their elaborate cloaks of electrons torn from them in the scrimmage. The lost electrons are speeding a hundred times faster to find new resting-places. Look out! There is nearly a collision as an electron approaches an atomic nucleus; but putting on speed it sweeps round it in a sharp curve. A thousand narrow shaves happen to the electron in 10^{-10} of a second; sometimes there is a side-slip at the curve, but the electron still goes on with increased or decreased energy. Then comes a worse slip than usual; the electron is fairly caught and attached to the atom, and its career of freedom is at an end. But only for an instant. Barely has the atom arranged the new scalp on its girdle when a quantum of aether waves runs into it. With a great explosion the electron is off again for further adventures. Elsewhere two of the atoms are meeting full tilt and rebounding, with further disaster to their scanty remains of vesture. . . . As we watch the scene we ask ourselves, Can this be the stately drama of stellar evolution? It is more like the jolly crockery-smashing turn of a music-hall . . . but it is all a question of time-scale. The motions

of the electrons are as harmonious as those of the stars but in a different scale of space and time, and the music of the spheres is being played on a keyboard 50 octaves higher . . .

What a beautiful description! But what makes it even more beautiful is that the protagonists are the atoms that now make us up. We are truly stardust.

But as far as these primordial atoms are concerned, stardust has not yet been manufactured. Our young star still contains merely hydrogen, helium, and a dash of the other light elements produced in the Big Bang: deuterium (the nucleus of heavy hydrogen) and lithium, the next lightest element after helium.

Over the course of less than a century after the star reaches hydrostatic equilibrium for the first time, all this will change. For as this massive star slowly contracts, its interior temperature increases much more quickly than a lower-mass star's would. In turn, the surface temperature rises, allowing radiation to transport heat more efficiently. In this way, the luminosity of the star stops decreasing as the star contracts. This is possible only if the temperature on the surface of the star increases. In turn, as more of the heat is transported by radiation, the huge convection pattern settles down, and ceases to dredge up material from deep inside the star to the surface, and subduct it from the surface to the core. More and more of the heat is transported by radiation and less and less by convection. The surface of the star continues to increase in temperature, and the inside of the star gets progressively hotter as well.

During this phase, from the outside the star appears very luminous but irregular. The huge convective flows cause the brightness to vary rapidly, and unevenly. Such an object is called a *T-Tauri star*, named after the first of a number of such stars observed with modern telescopes in gas clouds in the constellation Taurus, representing stars caught in the act of forming today.

Our 4 hydrogen nuclei in particular end their rollercoaster ride throughout the star as the convection flows shrink, and they settle somewhere midway between the surface and the center. By this point they are neighbors in an astronomical sense, although they are still separated by a macroscopic distance — several kilometers at least. Our helium nucleus finds itself nearer the surface, although as the star evolves, it sinks slightly inward because it is heavier than hydrogen. The density of matter in this region of the star is comparable to that of water. While this may not seem terribly exotic by terrestrial standards, the density vastly exceeds anything that has been seen in the universe for several hundred million years. The contraction over the course of a few hundred thousand years has increased the density of the primordial molecular cloud by a factor of over a billion billion. At this density, new physical processes can take place.

Deeper down inside the star, the temperature now exceeds 1 million degrees, and at this temperature the first nuclear reactions since the earliest moments of the Big Bang begin to occur. They provide more of a whimper than a bang, however. Hydrogen nuclei, namely protons, can collide with the weakly bound nuclei of deuterium (containing a proton plus a neutron), and stick together to form a more tightly bound rare isotope of helium, helium-3 (different isotopes of the same element have the same number of protons but differing numbers of neutrons). In the process the system radiates away the energy released as the particles bind together. Essentially all of the initial deuterium in the central regions of the star can disappear within a matter of years in this way, and in the very central regions in a matter of seconds after nuclear burning starts!

Unfortunately, the energy generated by this process merely slows the contraction of the star, and cannot stop it completely. This is because the initial amount of deuterium is so small that the total energy generated even if it is burned very quickly is too small to generate sufficient pressure to counter the gravitational attraction. Similar burning of other trace elements such as lithium at a temperature of a few million degrees does not generate enough energy to significantly alter the energy balance of the star.

We now face Eddington's conundrum head on. What can stop the further collapse of the star? If no additional energy source kicks in, the contraction will continue unabated, and after a million years or so, the star will have shrunk away to a dense, compact, cooling chunk of matter.

As the star continues to shrink, and the temperature of its core continues to rise, protons smash against each other with ever-greater energy. Yet because the protons are charged, even the collisions associated with a temperature of 10 million degrees are generally about 1,000 times too weak to get the protons close enough so their nuclear shells overlap.

If they could overlap, then a new nuclear reaction first recognized by Bethe in 1939 could occur. Two protons could collide to form a proton and a neutron. For such a reaction to take place, the weak force, described earlier, which can transform protons and neutrons into each other, would have to play a role. But the range over which the weak force operates is very small, smaller in fact than the size of the atomic nucleus. Thus unless the protons essentially strike each other head on, there is zero likelihood of such a transformation taking place.

There is, of course, another problem. Neutrons weigh more than protons. Thus the effect of changing 2 protons into a proton and a neutron is to suck energy out of the surroundings, and not generate energy which can then serve to hold up the star. But if the proton and neutron produced by the collision can bind together to form the nucleus of deuterium, heavy hydrogen, then the situation changes completely. Deuterium weighs less than the sum of the mass of 2 protons. In fact, the difference in mass implies that the energy released by turning 2 protons into a proton and neutron bound in deuterium is about 1 million times as great as the energy released in a typical chemical reaction.

But it doesn't stop here. As I have discussed, the deuterium nuclei can, in a matter of seconds after being produced, capture another proton to produce helium-3. The helium-3 nuclei can collide with deuterium, hydrogen, or with themselves. Initially, when they just begin to be produced, the likelihood is far greater of being destroyed by collisions with hydrogen or deuterium. Eventually, if their density can build up sufficiently, the dominant reaction of helium-3 will be with helium-3. During such a process a proton and neutron in one nucleus can pair up with a proton and neutron in the other nucleus, combining to form stable helium-4. In the process, 2 protons are released. Because helium-4 is very tightly bound, the total energy released in this process is almost 6 times as much as that released when a proton and neutron combine to form deuterium.

The net effect of these reactions is that 6 protons interact over time, to eventually produce helium, containing 2 protons and 2 neutrons, leaving 2 protons left over. Thus 4 protons have effectively converted into a helium nucleus, and the total energy released is 20 million times as much as one would achieve by burning an equivalent mass of coal. A process had finally been discovered that could keep the sun burning for billions of years! When the Mike bomb exploded in the Pacific Ocean, the awesome power of nuclear fusion became clear to the whole world, and not just a small cadre of nuclear physicists and astrophysicists. The secret power source of the stars had been manifestly exposed.

But there remains a problem. How can the whole thing get started? Remember that even at a temperature of 10 million degrees, and the densities appropriate at the core of the emerging star, not a single proton possesses enough energy to collide head on like a billiard ball with another proton and create a deuterium nucleus.

This is yet another time we can thank the gods of chance. First, the protons inside the sun are in a *thermal distribution*. This means that some protons can have more energy than the average proton. For example, about 1 proton in 10 million can have 10 times the energy of the average proton in this hot gas. Next, at the scale of atoms and nuclei, the laws of quantum mechanics reign supreme. Here, things that are otherwise impossible may be merely improbable. Thus while even the most energetic protons with any significant abundance inside the sun have 100 times too little energy to overcome the classical electronic repulsion between protons in order to participate in a nuclear reaction creating deuterium, quantum mechanics creates a tiny probability that these protons can nevertheless sneak a quick kiss and maybe more.

At the heart of quantum mechanics is the fact that particles such as protons do not behave like billiard balls. I can throw a billiard ball against a wall a million times and it will either bounce back or it will break through, but a proton can sometimes start out on one side of a barrier and end up on the other side without ever actually having to go through the middle. Unless the barrier is infinitely high, there is always a nonzero probability, albeit potentially very small, that the proton will be found at one instant on one side, and at the next instant on the other. So the electronic barrier between protons may be discouraging but never completely daunting. Every now and then protons can collide without actually having to fully overcome their mutual repulsion. The probabilities may be very small, but there are a lot of protons available in a star, and there are a lot of collisions occurring during the star's lifetime.

The result is a tender balance of opposing demands. As one increases the energy of a proton inside the sun, one increases the likelihood that in a collision with a partner it can "tunnel through" to penetrate the other particle. On the other hand, in a thermal gas the number of protons with a given energy quickly decreases as that energy begins to exceed the average.

The same rules apply to protons as to politicians. Being pulled both ways usually favors the choice of a middle ground. In the case of the conditions appropriate to the temperature when nuclear burning begins in earnest, about 15 million degrees Celsius, the protons favored to initiate nuclear burning carry about 13 times the average thermal energy of protons in the gas. About 1 in 100 million protons is fortunate enough to have this energy at any one time, and even then, such protons can collide with their neighbors for billions of years before successfully reacting to produce deuterium. Even so, the energy produced by the ensuing reactions that will ultimately yield helium is so great that the heat they generate can create a pressure that can completely counter further gravitational contraction of the star. Once these nuclear fusion reactions turn on, the star has a right to be called by that name.

In the spirit of Eddington, the inside of such a newborn star is not only plenty hot enough, if it were any hotter we would be in trouble. If the nuclear reactions were not so rare, the star would burn up all its nuclear fuel in the wink of an eye, ending its life just as it began, in a huge uncontrolled thermonuclear explosion. But instead, the star is self-regulating. If the core gets hotter, the pressure increases, causing the gas to expand against gravity, thus cooling the core. In this way, the chaotic, "hurly-burly" lifestyle of Eddington's stellar atoms gets converted, on large scales, to a remarkably stable object, with a lifetime that stretched the imagination of the greatest minds of the nineteenth century, and whose stability might ultimately nourish the slow evolution of life on nearby planets.

Not in the case of this star, however. Here, the marching orders of gravity cannot be ignored for long.

When the first thermonuclear explosion to take place on Earth occurred on that warm November morning in the Pacific, the dreams of centuries of alchemists became reality within a fraction of a second. From the moment the chemical explosive "fuse" was ignited to the time in which uncontrolled explosive fission and fusion processed any material in the immediate vicinity of the bomb core into something new, a bird on the island about to be vaporized would not have had time to flap its wings.

In the case of our massive star, just born, the nuclear fuse was far longer-burning, but the final result was just as inevitable. A bomb of unprecedented magnitude was destined to ultimately explode the moment the first fusion reaction to produce deuterium occurred inside its fiery core. The lives of our atoms would forever be changed as a result.

For the time being, however, life for our 4 hydrogen nuclei and 1 helium nucleus proceeds rather uneventfully. Located outside the dense inner core of the star, they merely get bombarded by their neighbors traveling at speeds of hundreds of kilometers per second, trillions of times per second, for perhaps 10 million years. During this time, in the absence of large convection flows, each of our atoms may travel perhaps a few kilometers within the star due to its own random thermal motion. In so doing, our 4 hydrogen atoms slowly converge. With a temperature less than 1 million degrees, however, they have no chance to participate in the nuclear burning reactions occurring deeper inside the star. Our helium nucleus has even less chance to party. No stable isotopes exist with mass number 5, so there is not any stable material that helium could form by capturing a hydrogen nucleus in any collision.

But deep inside the star, things are getting more exciting. After 10 million years of burning with a luminosity 10,000 times that of our sun, this hungry giant has burned almost all the available hydrogen in its core to helium. During this time, the pressure in the core is reduced as hydrogen nuclei get converted into helium nuclei. By virtue of their larger mass, they move more slowly than their predecessors. The core starts to slowly contract, getting hotter in response, and releasing energy to the rest of the star. This additional heat causes the outside of the star to puff up once again. This would serve to cool the core, except that the loss of pressure as helium is generated keeps driving the core to contract further, heating up even more.

In this way, the layers surrounding the core of the star heat up further, powered now by the contraction of the core, and also by the fact that as the temperature in these layers increases, hydrogen burning into helium begins in these layers. At this point, almost all of the luminosity of the star is now powered by the hydrogen fusion reactions in the shell surrounding the largely helium core. A significant fraction of the energy produced in the hydrogen shell is in fact absorbed by the outer layers of the star, which continue to expand in response.

This process continues for several hundred thousand years at full strength, but over the course of another half million years, as the core continues to contract and increase in temperature, the hydrogen-burning shell is unable to provide sufficient energy and the star begins to cool at its surface while its radius increases. The star begins to become redder.

This intermediate breather phase, a seventh-inning stretch if you will, is not the most exciting time in the star's life, but it takes on a profound significance for our hydrogen atoms. For it is during this 500,000-year breather that our 4 hydrogen atoms finally see some action. Two have now approached each other within an atomic radius, and after tens of thousands of years of jostling, they finally fuse to form deuterium. Like a fruit fly, within minutes of its creation our fledging deuterium nucleus ends its life, this time through a collision with another one of our hydrogen nuclei. Together, they form the nucleus of helium-3. Meanwhile, our fourth hydrogen atom, some distance away from the other 3, has participated in another series of fusion reactions leading to the production of a helium-3 nucleus.

These two helium-3 nuclei now wander hither and yon, colliding with hydrogen atoms over the course of this full period of stellar contraction, until they finally meet. They collide with just the right energy after billions and billions of collisions since their birth a few hundred thousand years before and in a flash produce a nucleus of helium-4, spitting out 2 protons in the process. What started as 5 nuclei — 4 of hydrogen, 1 of helium — have now become 2.

Deeper inside the sun, the inward contraction would continue unabated, except for the fact that nuclear fusion of hydrogen to helium is not the end of the line. At first glance it looks like it might be, and for a long time this presented a severe roadblock to those who tried to understand how all of the elements beyond helium that govern our existence might be formed. No sooner had physicists solved the remarkable problem of how to make nuclei that contain neutrons when one begins only with protons than they had to face yet another puzzle from the subatomic world. While production of helium can occur by the simple series of processes described here, there are no stable nuclei with mass number

5 or 8. With only hydrogen (mass number 1) and helium (mass number 4) having any significant abundance, there are no other alternative mass numbers that can result from collisions between either a hydrogen and a helium nucleus or between 2 helium nuclei. No chain of light-particle reactions seemed to offer a way to leap beyond mass number 8.

Yet the next two most abundant nuclei in nature today, after helium, are carbon-12 and oxygen-16, so somehow this hurdle had to be overcome. The fact that these two nuclei contain precisely the number of protons and neutrons contained in three helium-4 and four helium-4 nuclei respectively suggests that somehow helium reactions must be the key.

Yet the probability of 3 helium nuclei coming together at the same time is so small that fewer than 1 hundred-millionth of the helium atoms in a star could fuse in this way to form carbon during a typical stellar lifetime. More important, perhaps, is the fact that there is no way such three-body collisions could generate energy at a rate fast enough to combat the inevitable collapse of the stellar core driven by gravity.

The temperature in the contracting core of our star now exceeds 100 million degrees and counting. Two more accidents of nuclear physics save the day, or rather, continue to make daylight possible. First, when 2 helium nuclei come together, if they were to "stick" they would form the unstable nucleus of beryllium-8. This nucleus decays back again into 2 helium nuclei. However, the mass of beryllium-8 is only slightly more than the mass of 2 helium nuclei. As a result, there is barely enough energy to decay, causing the beryllium nucleus to live for almost 1 millionth of a billionth of a second. This may not seem like much, but it is about 100 million times longer than one would expect 2 helium nuclei to hang around together if they weren't momentarily connected in a larger nucleus. As a result, at any one time one expects about one beryllium-8 nucleus for each billion helium atoms in the dense core. One millionth of a billionth of a second is also long enough for some helium nuclei to wander by and collide with some of the beryllium nuclei before they decay, and stick to form the stable nucleus carbon-12, the building block of all organic materials.

Here the second nuclear miracle comes into play. With so few beryllium nuclei around to interact with, collisions are at a premium and every one should, if possible, count. This is not normally the case with random collisions. It was recognized early on by the astronomer Fred Hoyle, the man who came up with the term "Big Bang" (as a term of derision, actually, because he was pushing an alternate, "steady state" theory), that the only way to sufficiently enhance the probability of forming carbon in a nuclear reaction between helium and the unstable beryllium nucleus in order to power stars would be if an excited state of carbon could be created precisely in collisions that might occur in the energy range accessible in the interior of stars. Such a "resonant" reaction can occur with a probability hundreds of times greater than would otherwise be possible. Sure enough, subsequent work in the laboratory demonstrated just such a state in carbon-12. The pathway beyond helium had been discovered and a new life for the star begins!

The burning of helium to form carbon opens up a new phase for the universe. Once carbon has been formed, the gateway to creation of all the heavy elements that dominate our own existence on Earth is opened. With all the power of the Big Bang, the helium barrier could not be overcome. The birth, and subsequent expansion, of the universe was just too fast. Instead, slow and steady wins the race. Over the course of 10 million years, the slow contraction of the star, combined with the buildup of helium and the ever increasing temperature and density in its core, make possible rare reactions that require thousands, or millions, or even billions, of years to occur.

Still, this new energy source for the star is a poor substitute for hydrogen burning. Per unit mass, far less energy is released in forming carbon than in forming helium. Reactions must thus proceed at a much faster pace in order to generate the same pressure in the stellar interior. It may take 10 million years for the hydrogen in the star's core to be exhausted, but it takes merely 1 million years for the helium fuel to be fully used. In addition, the probability of bringing 3 helium atoms together to form carbon is such a strong function of temperature that only in the hottest part of the core can these reactions occur at any significant rate. During this time the bulk of the star's luminosity is still produced by hydrogen burning in the shell outside the core. Moreover, the increased heat of the core continues to cause the exterior of the star to expand.

Lesser-mass stars sometimes barely survive the beginning of helium burning. Because it is such a sensitive function of temperature, the turn-on of this process can be explosive. The new energy generated by helium burning heats the core, which in turn results in more rapid helium burning, and so on. The "helium flash" almost blows the star apart. A star as massive as this first-generation behemoth make the transition much more smoothly, however.

After a million years, the helium fuel in the core is now significantly reduced. The core contracts further and heats up still more. Eventually the temperature becomes high enough for a helium nucleus to collide with a carbon nucleus — containing 6 protons with a large enough charge to repel all but the most energetic helium nuclei — and form oxygen-16, the dominant isotope of oxygen in the universe. Even though the repulsion of carbon for helium is great, the fact that only two bodies need to collide to form oxygen means that as the helium abundance begins to drop in the core while the carbon abundance begins to increase, the remaining helium atoms preferentially collide with carbon to form oxygen. Thus by the time helium is exhausted in the core, significant quantities of both carbon and oxygen have been created.

By now, two shells exist outside of the core. A hydrogen-burning shell (in which our 2 helium atoms were formed a million years earlier) surrounds a helium-burning shell, which surrounds the carbon–oxygen core.

Things are really beginning to become desperate for the star, however. With helium exhausted in the core, contraction proceeds once again. In order to produce new energy, carbon must interact with carbon. But the charge barrier is larger now, and the star must heat up to about 600 million degrees before this process can begin at a significant rate. Carbon–carbon collisions can produce a plethora of nuclei, from oxygen to sodium to magnesium. Carbon burning releases less energy than helium burning, and once again a more rapid reaction rate must be maintained to fight the inexorable march of gravity. Carbon burning lasts a mere 100,000 years, 10 times shorter than the phase of helium burning in the core, which is in turn 10 times shorter than the phase of hydrogen burning.

Things are now rapidly getting out of control. Oxygen is next, burning to produce silicon, again an element of vital importance for Earth-like planets, and also sulfur at yet higher temperatures. But oxygen burning proceeds for merely 10,000 years, again a factor of 10 less than its predecessor fuel.

Now temperatures are so hot, approaching 1 billion degrees, that the radiation itself is energetic enough to break apart nuclei formed earlier. Thus, for example, neon, formed when oxygen captures a helium nucleus, can break apart back into oxygen and helium. The helium can now be captured by remaining neon nuclei, which in turn produces magnesium. In so doing, the oxygen abundance can increase further in the shells surrounding the hottest regions of the core. Oxygen is the third most abundant nucleus in nature, following hydrogen and helium, and this is the reason why.

Once silicon burning begins, the core of the star is desperate for energy. Every fusion reaction subsequent to helium releases less and less energy per unit mass. Moreover, photo-destruction of nuclei takes energy! Of course, after a proton or helium nucleus is dislodged, and subsequently captured by another nucleus, net energy can be released.

One can still ask, How long can fusion keep producing energy? After all, the large positive electric charge in the nucleus will eventually win out over the nuclear attraction. As with many aspects of the universe, the ultimate configuration of matter depends on a very simple scaling law. As one proceeds from helium to carbon to oxygen to silicon, the amount by which each proton and neutron in the nucleus is bound to the nucleus increases. This process continues until one reaches iron. Iron-56 is the most tightly bound nucleus in nature. After iron, all heavier nuclei are less tightly bound. As I described for neutrons bound in nuclei, relativity tells us that bound objects weigh less than unbound objects, because the bound objects have less total energy (that is, it takes energy to unbind them). This simple result tells us that we can always gain energy by fusing 2 light nuclei to make a heavier nucleus, until we get to iron. Once we reach iron, the addition of any further protons or nucleons will create nuclei heavier than the sum total mass of the particles that go into the mix. Making such heavier nuclei will then take energy, rather than release it.

For this reason, all stars are doomed eventually to run out of nuclear fuel. And once our star starts to burn silicon, it is dangerously close to the end. In principle, silicon can burn with silicon to directly produce iron, but in practice things are too hot for this to occur directly. Instead, silicon and the other elements present in the dense core are knocked apart by radiation, only to reassemble in new configurations. As long as the nuclei that are reassembled are more massive than the ones that split, this will release energy, until iron is produced. Then one is out of luck.

Incredibly, the journey of nuclear burning from silicon to iron in the core of our star lasts but a single day! For 10 million years all of the nuclear reactions holding the star up against gravitational collapse have been leading to this single last gasp. Almost 10 million years of hydrogen burning, followed by 1 million years of helium burning, 100,000 years of carbon, 10,000 years of oxygen, and then a single day for the rest of the trip. Once it is over, there is no hope. In fact, the dense inner core of the star, now surrounded like an onion by shells of oxygen, carbon, helium, and hydrogen, is about to undergo one of the most traumatic events in all of the visible universe. Our 2 helium atoms, far removed from the inner action, will nevertheless suffer the consequences.

9.
A PRETTY
BIG BANG

Anybody who is not shocked by this
subject has failed to understand it.
NIELS BOHR, SPEAKING ABOUT
QUANTUM MECHANICS

In the cold mountains of Chile, where the air is thin and the nights are crisp and clear, a meandering road climbs up to a cluster of modern cathedral-like domed buildings. There, high above the clouds that obscure the valleys below, giant machines of the night are at work, searching for signs of death in the sky.

At the Cerro Tololo Inter-American Observatory near La Serena, Chile, an international collaboration of astronomers has mastered the art of probability in a high-stakes competition to determine what the universe is doing. Once every 100 years or so in a galaxy containing perhaps 100 billion stars, a single star explodes in a fireworks display unparalleled in the universe. Remarkably, in our own Milky Way galaxy such events often go unnoticed, because we cannot see the forest for the trees. Most of our galaxy is obscured to our view because it is full of dust, which absorbs visible light. As a result, we can often see more clearly what is happening in galaxies millions of light-years away than we can in our own backyard.

Such observations have, however, been inscribed in human history. On the twenty-second day of the seventh moon of the first year of the period Chih-ho (August 1054, during the reign of the Emperor Ren Zhong) in China, the Chief Calendrical Computer of Imperial China, Yang Wei-De, reported the following:

> Prostrating myself, I have observed the appearance of a guest-star in the constellation T'ien Kuan; on the star there was a slightly iridescent yellow color. Respectfully, according to the dispositions for Emperors, I have prognosticated, and the result said: The guest-star does not infringe upon Aldebaran; this shows that a Plentiful One is Lord, and that the country has a Great Worthy. I request that this prognostication be given to the Bureau of Historiography to be preserved.

Wei-De's prophecy may have merely flattered the emperor, and perhaps caused him to rule with a gentler hand for some time. But human history is fleeting. The object first recorded in 1054 has stayed with us ever since, and is now known as the Crab Nebula. A beautiful sight to see through a telescope, it still shines with the light of greater than 75,000 suns, almost a millennium after Wei-De. It is an interesting sociological phenomenon that in 1054 the Crab "guest-star" would have been visible day and night for weeks in Europe as well, but there is no record in the writings of this time of its having been seen there. This period in European history is not called the Dark Ages for nothing!

More significant for subsequent human history was the observation of a supernova in 1572 by the Danish astronomer Tycho Brahe. This so impressed King Frederick II of Denmark that he provided Brahe with an island from which to do observations of the heavens. On the isle of Hven, now part of Sweden, Brahe spent 20 years observing the motions of the planets without the aid of a telescope, this instrument having not yet been invented. Nevertheless, Brahe's measurements were so accurate that they were later used by Johannes Kepler to derive his three laws of planetary motion, which were in turn used by Isaac Newton to derive his universal law of gravity. These developments revolutionized the modern world. Not only did they lay the basis of the modern science of physics and astronomy, they also suggested that the entire universe might be explicable in terms of natural laws. It has been argued that the burning of witches ended in Europe partly in response to the recognition that all effects could have natural, rather than supernatural, causes.

So a single supernova seen in the right place at the right time did noticeably alter the course of human history. Beyond this, other supernovae, long, long ago, and far, far away, made human history possible in
the first place. Let us return to this ancient time.

As far as our 2 helium atoms are concerned, all is well. They are im-
mersed in a dense gas of hydrogen atoms, and the temperature in this
middle region of the star still exceeds 10 million degrees, so that fusion
reactions of hydrogen are continuing to produce energy that makes its
way out to the surface. However, on this, the last hour of the last day of
the life of this star, silicon nuclei deep in the core have been fusing to
form iron at a fast and furious rate. The bulk of the star's core, with a
mass exceeding the mass of our entire sun, and a radius larger than the
size of our Earth, has already converted to iron, and there is no place left
to go. Within a second, it will all be over.

As the silicon burning ends, the core pressure drops and it begins to
contract. But now, the temperature rises to over 5 billion degrees, and
the radiation energy is so intense that all of the work of the past 10 million years is undone. Photons are so energetic that they break apart the
iron nuclei back into helium. This takes energy, rather than releases it,
and it sucks the thermal energy, providing pressure, from the core.

Things now begin to run away. The core begins to collapse much
more quickly. As it does so, its density continues to increase. When the
density reaches about 10,000 tons per cubic centimeter (!), the electrons,
which are being squeezed together along with the nuclei, gain enough
energy to convert protons back into neutrons, sucking more energy out
of the collapsing system. What is worse, in the conversion process neutrinos are emitted, and these neutrinos are so weakly interacting that
they escape from the star at the speed of light. As this energy is removed,
the core collapses even faster. But as it collapses, it becomes denser still,
causing more electrons to collide with protons to produce neutrons, taking more energy from the core, and so on. The core implodes with more
force than it is possible to picture.

Imagine an object the size of the Earth collapsing to form an object
the size of Manhattan in less than 1 second! My mind can never do this

justice. Nevertheless, the universe is not constrained by the limits of human imagination, and this is precisely what happens to the inside of our star.

So vigorous a collapse is hard to halt, but again nature manages the feat, although at a cost. As the density of the inner core reaches 100 million tons per cubic centimeter, what were the nuclei of atoms, and are now predominantly neutrons, are packed so closely together that they touch. The core becomes, in essence, a giant atomic nucleus, a physical realization of the configuration we simply imagined in chapter 1. At this instant, a new repulsive force sets in. The neutrons cannot overlap with each other. The laws of quantum mechanics allow only a certain number to be squeezed into a small region. This new nuclear repulsive force is so great that even this immense inward collapse cannot proceed. Instead, the inside of the star — now known as a *proto-neutron star* — "bounces."

Whoever first spoke of getting stuck between a rock and a hard place had no idea how much worse it can really get. Devoid of pressure, the densest region of the core collapses in a second. The outer shell of the core, still largely iron, suddenly finds itself without anything to support it, and begins racing toward the center of the dying star. As this shell collapses, material falls inward at more than 60,000 kilometers per second, about 20 percent of the speed of light! Then, *boom*, it hits a metaphorical brick wall, except that this wall is 100 billion times harder than brick. Like a baseball being hit by a bat, the material at the interface is sent shooting outward again. A dense shock wave, traveling about 10,000 kilometers per second, plows through the infalling material. At this speed it would take about 30 minutes to reach the surface of the star. Smashing into the rain of infalling material, however, it quickly loses energy and slows almost to a halt.

Meanwhile, most of the rest of the star, containing our 2 atoms, among others, is blissfully unaware that anything has happened. In 1 second, the slowing pressure wave communicating the state of the interior to the outside can travel at most a few thousand kilometers. Like the cartoon character Wile E. Coyote, who remains suspended momentarily after walking, driving, or hopping off a cliff, the rest of the star, extending out millions of kilometers, is not yet aware that it is supposed to fall.

In fact, the rest of the star will never know. Before it can collapse, one of the most remarkable series of events in nature occurs, in about 10 seconds. First, just behind the stalling shock wave, the density of material is much higher than its surroundings, and material keeps piling up on this surface. The pressure of this high-density material is dominated by the energetic electrons that are released as protons convert to neutrons and is large enough to stop the shock wave from turning inward.

Before this bounce, the neutrinos produced by the conversion of protons to neutrons in the initial collapse have escaped, in a hundredth of a second or so, as neutrinos are wont to do. After all, the average neutrino emitted in such a process can travel through several thousand light-years of lead before interacting even once! Solar neutrinos, for example, produced by reactions inside the core of our sun, go right through the Earth without knowing it is there.

In the dense collapsed core of our star, however, containing slightly more than 1 solar mass of material confined to a region of 50 kilometers or so in radius (it has bounced outward from its initial collapse radius of 10 kilometers), the environment is unlike anywhere else in the universe. With a mass in excess of 1 million Earths confined to a region the size of a small city, and with a mass of Manhattan contained in each cubic centimeter of material, the density is so great that not even neutrinos can escape from the inferno. And there are neutrinos in abundance! At a temperature of about 5 billion degrees, the interior of the proto-neutron star emits neutrinos as freely as photons. These neutrinos build up in abundance because they become trapped inside the star. Remember that it takes 100,000 years for trapped photons to escape from the sun instead of the 2 seconds it would take them if they headed straight out, without collisions, at the speed of light. So, too, neutrinos emitted inside the core of the proto-neutron star take on the order of 10 seconds to escape, over 1 million times longer than it would take them to emerge if they did not scatter on their way out.

These neutrinos continue to build up in the dense, hot region behind the shock front, and they have two effects. First, they slow the transition of protons to neutrons in and around the core. As their presence builds up, they can interact with neutrons, converting them back into protons. Thus, until the neutrinos radiate out of the core region, the core cannot become a neutron star. More important, however, is the fact that the neutrino buildup increases the pressure behind the stalled shock front, just as bubbles of gas in a viscous liquid can lift up its surface. Slowly at first, the shock wave moves outward, increasing its speed as the infall slows and it encounters ever more diffuse material. Within minutes, the shock wave reaches our 2 helium atoms, and carries them as a wave carries a surfer, toward the surface, and beyond, as the star explodes and, in its last gasp, lights up the night sky with the light of a billion suns.

Our atoms are free at last, but they are not alone. As the surface material shoots outward at tens of thousands of kilometers per second, the star emits as much energy in a few weeks as our sun has over the past 4.5 billion years. Moreover, the shock wave dredged up material from throughout the star. Next to helium, oxygen is the second most abundant fusion product from stars, followed by carbon, then nitrogen, neon, silicon, magnesium, sulfur, and iron. All of these materials from the shells existing outside the inner core are expelled, along with helium and hydrogen. In the hot, expanding, neutron-rich region behind the shock wave, capture of neutrons by intermediate elements quickly produces all the elements up to uranium, containing a total of 238 protons and neutrons.

This material smashes into the interstellar material that surrounds the dying star. In the process, this gas is heated to temperatures in the neighborhood of 1 million degrees. As the shocked material propagates outward, the material both in front of the shock front and behind it cools. If the material surrounding the supernova is not too dense, the ejecta can travel significant distances before cooling. If it is denser, then following its compression this material will radiate away its energy, and slow down more quickly. Barring collisions with extremely dense gas clouds, the bubble of ejecta gas will continue its outward expansion at thousands of kilometers per second for hundreds or thousands of years. After this time it will slow to merely hundreds of kilometers per second. It will take almost 100,000 years for the remnants to fully dissolve into the background interstellar medium.

In this hot, energetic environment, which is full of heavier elements, complex chemistry can finally for the first time occur. Nuclei will cap-

ture electrons and reduce their state of ionization, and collisions between atoms will transfer electrons. As the material cools over the course
of months, years, and millennia, the heavier elements such as iron will
condense into microscopic solid grains, on the surface of which other
materials may collect.

Our 2 helium atoms can remain ionized for some time in this hot,
dense environment. But helium is a noble gas, that is, compounds containing helium do not readily form. Nevertheless, our helium atoms play
an important role in the chemistry of this expanding cloud. The density
of matter is still high, perhaps 1 million times higher than the density of
the gaseous nebula into which the shock material is expanding. Helium
and the next dominant elements, oxygen and carbon, can sometimes experience a three-body collision, which can cause the latter two elements
to react to form carbon monoxide, some of which survives as the gas
cools. Carbon monoxide will later play an important role in the lives of
our atom. In a massive star, such as the one that has just exploded, more
oxygen than carbon is produced, so the oxygen that is left over after carbon monoxide eventually forms combines with iron, silicon, and hydrogen to form iron oxides, silicates, and water.

As the gas bubble expands, each time it encounters a region of gas it
compresses it. If the region is sufficiently dense, once compressed, collisions will occur fast enough to cool the gas, and it will go from millions
of degrees to thousands of degrees, emitting light in visible wavelengths.
Thus, as the shock wave passes, filamentary gas clumps will light up in
turn along its route, like Christmas tree lights alternating on and off.

Four of our original hydrogen atoms live in one such clump of gas, located about 20 light-years from the site of the supernova explosion. After
about 5,000 years, the expanding shock wave reaches this region, and the
atomic collisions compress the gas in the clump, and at the same time
some of the material from the expanding gas bubble trails behind with
the gas clump, which is imparted a velocity of about 100 kilometers per
second. The compression wave causes the gas to release a burst of heat
and light, cooling to a temperature of 1,000 degrees Celsius in the course
of years, not millennia. At the same time, the high density of gas and dust
grains surrounding this region shields the material inside from the radiation of the supernova, and other nearby stars, allowing it to remain
cooler than the background hot gas.

Our 2 helium atoms now experience déjà vu all over again, as they and the 4 hydrogen atoms have become a part of a new stellar nursery. By the time the temperature falls below a few hundred degrees above absolute zero, the active chemical reactions cooking the gas and dust debris from the supernova slow to a halt. When the dust has settled, literally, the new environment is superficially the same as the first molecular cloud our initial 5 atoms became a part of, but in detail it is fundamentally distinct. Molecular hydrogen is still the dominant material around, but now there is a small pollution of other debris. The dust grains, onto which molecules of water vapor can solidify as the gas cools, actively absorb light from the surrounding stars and hot gas, and re-emit this radiation in much longer wavelengths of infrared light.

The carbon, then present at less than 1 part per 10,000, will play a crucial role in cooling the gas cloud in preparation for star formation. At early stages in the collapse of the cloud, carbon atoms act as an excellent refrigerator. Radiation from outside the cloud can knock electrons out of atoms, ionizing them. If the energy stored in the electrons gets transferred to the atoms, this will heat up the gas. The electrons, however, can collide with the carbon atoms, which can be excited at temperatures of about 100 degrees above absolute zero (173 degrees below zero on the Celsius scale). These carbon atoms will then de-excite by emitting radiation, which escapes the cloud. In this way, carbon, even present at very small levels, can keep the gas at 100 degree temperatures or less. Converting thermal energy to radiant energy that can escape the cloud will keep the cloud cool enough to begin to collapse.

The collapse into a new protostar proceeds much as before, except with a few new and significant complications. Heavy elements can be ionized more easily, and electric currents can flow, both creating and responding to magnetic fields, complicating the net collapse process. More important, as the core of the protostar collapses, the new dust surrounding the core plays an integral role. It will absorb and re-radiate radiation emitted by the collapsing core, shielding it from the outside environment. Even more important, the angular momentum of the cloud as it collapses can be carried outward by congealed grains, eventually coalescing to rocks and planetesimals, that will remain in orbit in the outer parts of the collapsing pre-stellar nebula. Material falling in toward the center of the cloud with a net rotation will collide and coalesce along a central disk of material that will orbit the collapsing gas sphere. The fodder for rocky planets to form around our collapsing star now exists.

The protostar that is now forming will eventually collapse into a star much like our own sun. Four planets will form around it, three giants like our own Jupiter, and one rocky planet, located just far enough from the star so that liquid water can exist on its surface. The present environment is horrendous, with intense radiation from the collapsing core bombarding every object in the surrounding disk, and grains colliding to form larger and larger objects, which collide with great intensity to form larger objects still. Yet 5 billion years later, any direct evidence of this early chaotic jumble will have been erased. A stable star will bathe the system in a constant warm glow, small meteors will have been largely ejected from the system by the gravity of the large planets, or will have already collided with one of these large objects billions of years earlier. Only intelligent beings, capable of exploring the remnants in this solar system and of deducing the past from remote clues in the present, might have a hope of unraveling the details of the cosmic drama that led to its creation.

But we will never know if life formed around this star, which has by the present day, some 10 billion years after these events, exhausted its hydrogen fuel, and grown in size to gobble up the once hospitable inner planets. For our 4 hydrogen atoms — now 2 hydrogen molecules — and 2 helium atoms escape the evolving inferno. As material falls inward toward the core of the collapsing cloud, intense magnetic fields combine with complex dynamics of the rotating system to cause two jets of material to fly outward from the north and south poles of the rotating spherical protostar at the center. Like hot water shooting out of a steam cleaner, these energetic jets bore holes in the surrounding dust and gas, spewing material back into the interstellar medium, away from the collapsing star. Caught up in this astrophysical whirlpool, our atoms get shot out into the emptiness of space once more.

This respite does not last long, however. The massive cloud that originally fragmented to collapse into our first massive star contains the raw material sufficient to make more than 1,000. Supernovae blow some gas completely out of the surrounding clouds, indeed completely out of the emerging galaxy, but they also trigger star formation in nearby regions as they compress gas and dump new raw materials into the brew. Within a million years, our 6 hapless wanderers find themselves deep inside the core of yet another new, much larger, collapsing cloud. This time they will not escape.

The environment our helium atoms experience is not new for them. They have been through all this before. Our hydrogen atoms have thus far avoided the fusion furnace, but not for long. Within a million years, the massive star they have become engulfed in begins to shine with the energy of fusion. The temperatures deep inside the star, where all 6 of our atoms, 4 of hydrogen and both helium atoms, now reside, exceeds 20 million degrees, and the cosmic cooking begins again.

For 100,000 years our hydrogen atoms survive the intense bombardment of radiation, but inevitably, they will fuse together to form yet another nucleus of helium. The process by which they fuse, however, is quite different from that experienced by their hydrogen cousins that fused previously to form our two helium stellar veterans, even if the end result is the same. In this new, second-generation star, the heavy elements carbon, nitrogen, and oxygen all exist in trace amounts, having been spewed out of previous stellar explosions. These elements allow a new cyclical pathway for helium formation. One of our protons joins with carbon-12 to form nitrogen-13, which decays (with a proton converting to a neutron) into a new isotope of carbon, carbon-13. Two more protons collide and attach, yielding oxygen-15, which decays (with another proton converting to a neutron) into nitrogen-15. Following a collision with our final proton, the bloated nitrogen nucleus kicks out a helium nucleus, decaying back again to carbon-12. This cycle continues sporadically in the emerging star, but as it heats up to temperatures in excess of 20 million degrees, it happens more frequently. Once the first of our protons gets caught up in the process, the completed helium nucleus is created within a day.

We now have 3 helium nuclei located inside this giant star, jostled by radiation, slowly diffusing closer and closer together. For another million years, the temperature remains too low for anything much to happen. Then slowly, as all the hydrogen in the core of the star is slowly turned to helium, the core begins to contract again, heating up more and more.

One of our nuclei collides with another helium atom to form beryllium, but before you can say "Rumpelstiltskin," the beryllium collides with a proton, and fissions back into its helium nuclear components in less than a billionth of a billionth of a second.

Ten thousand years pass, with each second heralding billions of new collisions. Slowly, yet inexorably, our helium nuclei drift together toward their ultimate fate. Two of our nuclei once again merge to form beryllium, but this time, before it can be shattered apart, the third collides with the unstable beryllium nucleus and, *wham*, a trembling carbon nucleus is formed.

The new nucleus is not yet out of the woods. What forms is an excited state, called a resonance, which means it has not long to live. It can lose energy in one of several ways. In one of these modes, our helium nuclei would be regurgitated back into the fray. In another, an energetic photon is shot out into the star, carrying part of the energy and pressure that will be needed to help hold up the core against collapse for thousands of years. Our nucleus takes this second route. A hundred thousand years later, the energy emitted by the radiating carbon nucleus will emerge from the surface of the star as visible light. About an hour after that it is absorbed by an ice-covered dust grain, which subsequently re-emits infrared radiation that escapes the local gas cloud and the emerging galaxy, and travels for billions of years through empty space.

The absorption of light on the dust grain is the most intrusive event that has occurred to this object in its brief history, but not for long. By the time the light emitted from our carbon atom deep inside the giant star reaches the surface, the processes which will lead to this star's demise are well under way. Further in, deeper toward the core, other carbon atoms have fused to form oxygen, then oxygen to silicon, and silicon to iron. Once again the fuse has been lit for a stellar explosion, and within minutes the shock wave that engulfs the outer layers of the star's core, including the region containing our carbon nucleus, emerges from the star to engulf the dust grain and both the dust grain and our carbon nucleus are spit out into the void of space.

Finally, 500 million years after the Big Bang, 8 of our initial protons and 1 nucleus of helium have fused to form carbon, the building block of all organic compounds. Because of its particular chemical structure, carbon can form multiple bonds, either with other carbon atoms to make long, stable chains, or with greedy oxygen atoms that like to hoard electrons and bind with carbon, "oxidizing" the molecule, or with hydrogen atoms, which are happy to donate their one electron to the emerging structure, "reducing" any positive charge therein.

The range of chemical compounds that carbon can form is virtually limitless, and in a sufficiently dense environment it can be reduced or oxidized to make them all. Ultimately, when combined with the other abundant species, hydrogen, oxygen, and nitrogen emerging from supernovae, carbon-based molecules can form self-reproducing structures that may one day change the way the universe itself evolves.

In the expanding dust bubble, our carbon atom joins with one of the slightly more abundant oxygen atoms created in the explosion, forming carbon monoxide. Ten such molecules exist in the gas cloud for every million or so hydrogen atoms. Nevertheless, carbon monoxide and water represent the dominant molecular components in the gas, next to hydrogen. As the new molecule streams out of the emerging gas bubble, two competing processes affect its future. Small dust grains of iron or silicates attract molecules to their surface, so that they become coated with an icy mantle containing solid ice, carbon monoxide, nitrogen, and other molecules. At the same time, intense radiation from the dying star, and later from nascent protostars, bathe these grains with energy, which can both knock material from their surfaces and cause chemical reactions to occur between the different constituent molecules therein.

This continual process of vaporization and condensation on grains, followed by photo-induced reactions, allows a complicated new form of chemistry to occur. Our carbon atom gets bound up first in carbon monoxide, CO; then carbon dioxide, CO_2; then methanol, CH_3OH; then ethanol, CH_3CH_2OH; and so on.

As the gas cloud interacts with the surrounding medium and cools, another familiar process occurs, as the dust grain containing our carbon atom is incorporated into a molecular cloud whose outer surface absorbs and re-emits radiation from the surrounding stars and novae so that the interior can continue to cool.

Once again a star begins to evolve. As the molecular cloud cools and collapses, the new energy source growing at its core begins to power a new set of chemical reactions. The dust cloud surrounding the nascent star absorbs radiation from inside as well as outside. As the emerging star grows in luminosity, first slowly, then sharply in its by now familiar turbulent formative T-Tauri stage, most of the dust is vaporized, and our atom, previously bound up in a complex carbon compound, is vaporized once again to a carbon monoxide molecule.

The gas atoms collide, losing energy by radiation, and accreting onto a disk surrounding the protostar. At the inner radius of the disk the temperature can exceed 1,000 degrees, while at the outside the temperature decreases to merely 30 degrees above absolute zero (on the Kelvin temperature scale), almost 400 degrees below zero Fahrenheit. Our carbon monoxide molecule, lying in the outer region of the disk, becomes bound to a dust grain, and again begins to participate in chemistry, this time powered by the radiation of its new host star. In this case, our carbon molecule reacts to form formaldehyde (CH_2O), which then reacts with ammonia and other nitrogen-bearing compounds on the dust surface until it finds itself a part of the structure NH_2CH_2COOH. Staring at the chemical formula for this compound may not be enlightening, but the name it has been given, glycine, may ring a bell. This is the lowest-carbon-number amino acid associated with self-reproducing organic life.

This remarkable structure would not normally survive the turbulent future to follow. However, our dust grain will, over the course of the next million years, collide with other grains, clumping together in the outer reaches of this newly forming disk system to build up larger and larger chunks of material. Shielded from the harsh external environment by the surrounding material, our carbon atom remains safely frozen in place, while all around it, a solar system slowly forms.

In the highlands of Antarctica near the same South Pole research station that is probing the primordial density fluctuations generated in the Big Bang, another group of researchers is looking not upward, but down at the surface of the miles-high ice cap that covers the frozen continent. The pristine ice surface provides a delicate burial ground for extraterrestrial visitors.

I do not speak here of X-*Files*–type aliens, but rather rocks of a strange hue, primarily iron, but containing conglomerations of small carbon *chondrules* — millimeter-size spherules embedded in the stone, which are presumably remnants of condensed drops of original material from the dust nebula that surrounded our nascent sun. As the ice surface in Antarctica melts, meteorite fragments emerge on the surface and stick out like a sore thumb. Hardy parka-clad geologists on snowmobiles roam the icy plains, harvesting these rocks as a lobsterman harvests lobsters.

Meteorites recovered from Antarctica are revealing exciting new aspects of the history of our solar system and the terrestrial planets therein. The carbonaceous chondrite meteorites are among the most primitive objects in our solar system, emerging from the asteroid belt halfway between the Earth and Jupiter. The abundance of nonvolatile elements (elements that are not easily evaporated into the surrounding gas) in these meteorites closely matches the abundance of elements in the sun. This suggests that these objects have not taken part in any significant chemical or physical transformations since the early solar system formed.

Mixed in amid the inorganic crystals in these meteorites, more than 50 different amino acids have been uncovered. On Earth, all amino acids involved in biological processes have one "handedness." That is, compounds with the same chemical composition can exist in a number of different configurations. Two equivalent configurations might exist which are mirror images of each other. One of these can be classified as "right-handed" and the other "left-handed." In meteorite samples, the amino acids retrieved exist with both left- and right-handed configurations, although not in the same numbers, as we shall later discuss, but nevertheless suggesting their extraterrestrial origin.

It is impossible for organic materials to survive the shock of impact with the Earth's atmosphere if they enter it with a velocity greater than about 10 kilometers meters per second, which corresponds to the escape velocity from the Earth. Objects of extraterrestrial origin characteristically have velocities relative to the Earth far in excess of this value. The only way for an object to slow down sufficiently before plowing full speed into the depths of the Earth's atmosphere is if it is so small that it is slowed in the tenuous outer atmosphere first. Small objects, less than about 100 meters in radius, fit the bill. Objects at the upper end of this range, however, will generate huge impact craters when they hit the Earth, which in turn generate great heat, which is likely to destroy any organic material present. Smaller objects, such as the tiny meteorites recovered in Antarctica, and even much smaller interstellar dust particles ranging from 1 to 100 millionth of a meter in diameter may, however, survive the journey to deliver such material safely to the terrestrial surface.

Deeper inside the disk of the emerging system than our icy grain is located, rocks are colliding to form planetesimals, and planetesimals to form planets. Recent apocalyptic blockbuster films, displaying graphically the impact of a large asteroid hitting the Earth's surface, give some idea of the violence of the processes involved. The environment is far too harsh for the formation or survival of life. As planetesimals collide to form planets, the impacts completely melt the participants, forming and re-forming planetary surfaces. Even as collisions abate, the radiation from the evolving star bombards the planets with stellar winds and ultraviolet radiation.

In this emerging solar system, a planet is forming at a distance where liquid water can exist. As the planet forms, water vapor present in the surrounding gas presumably is adsorbed onto the grains that grow to rocks that get incorporated into the emerging planet's surface. Any atmosphere that might be captured around the emerging planet is quickly lost, however. Large rocks and planetesimals continually bombard the growing planet, which incorporates them to build up a planetary embryo within 10 million years. This is precisely the time period over which the host star goes through its hot convective T-Tauri stage. When the host star's luminosity grows, the resulting violent stellar wind that emerges blows off most of the gas in the inner solar system, including any that had accreted to surround the emerging planets.

Where does the gas come from that eventually forms the atmosphere of this new planet? Farther out in the emerging solar system, ice-covered grains have been growing. At the distance where water can first condense to ice, about 5 times as far from the new star as the new terrestrial-like planet is located, grains collide and build up. Within 10 million years, a giant planet has formed, accumulating ice and gas. This planet clears up all the gas and dust in this region of the nebula surrounding the star. Equally important, as it grows its gravitational effect on surrounding material increases in importance. Icy grains much farther out get perturbed by this emerging planet, and are kicked into orbits that bring them to the outskirts of the solar system, and also into its interior. As comets, they cut a shiny path across the sky as they approach their host star. Over the course of 100 million years or so, billions of these objects from the frozen outer reaches of the solar system bombard the tiny inner planet, bringing water, carbon dioxide gas, nitrogen, and organic materials containing carbon.

Our carbon atom is on one such comet. As it smashes through the emerging atmosphere of the new planet, tremendous heat is generated. Much of the gas is lost well before the object collides with the planet. Our atom, however, makes it to the surface, where a tremendous explosion upon impact ejects it high into the sky, destroying the fragile complex organic compound it was in. The heat generated breaks apart molecules, spewing out carbon atoms, oxgyen, and the rest. Our carbon atom emerges as carbon dioxide gas, which slowly builds up a thick layer surrounding this new planet.

The heat of the impact releases the water in the comet as steam. Water vapor combined with carbon dioxide now covers the planet, which is slowly cooling, following 100 million years of impacts, many of which would have been sufficient to melt the planet's rocky surface.

As the planet slowly cools, water, released both in the comet impacts and as steam from the continually reheated rocky surface, condenses, and a blue ocean begins to cover much of the surface of this new world.

By now, after 100 million years or so, the host star has settled down to its long, slow stage of nuclear burning. The wild childhood and adolescence of collapse, turbulent convection, and rampant heat release have subsided. The star is now only 70 percent as bright as our own sun is today. With this decrease in brightness, our planet could easily cool to be an icy, barren wasteland. But the rich sheath of carbon dioxide now surrounding it protects it from such a fate. Solar radiation reaches the ground and is re-radiated outward as infrared radiation, which is trapped by the carbon dioxide inside the atmosphere, heating it up. Instead of cold and dry, this new world becomes hot and damp. A humid summer day in Houston would be nothing compared to this.

And so it would remain, were it not for another chemical miracle. Carbon dioxide with pressures as high as 10,000 times that of the current carbon dioxide content of our own atmosphere is readily soluble in rainwater. Forming an acid, H_2CO_3, the rainwater attacks rocks, forming carbonates, such as limestone, and silicates. This material falls to the ocean floors and is removed from circulation. In this way, over time, the carbon dioxide in the nascent atmosphere is steadily reduced.

The carbon that is buried in the newly forming ocean floors would eventually build up to the point at which no more carbon could be added, were it not for another fortunate circumstance. The ocean floors of this hot young planet are floating on a molten sea of rock. Convection, like eddies in oatmeal, churn up material, causing rocks from the surface to crash together, pushing some down into the interior and bringing fresh new material to the surface. The carbon that is so buried at the ocean floor is then *subducted*, as geologists put it, into the interior. In this way, carbon dioxide is slowly removed from the atmosphere.

Our new planet is well on its way to becoming a blue ocean world. Our carbon atom, however, has by now disappeared under the planet's surface. Within 300 million years of cooling, our carbon atom has become bound up in limestone and has been subducted into the planet's interior. There, as it slowly marches inward over the course of another 50 million years, the rock housing it is heated up, and our carbon atom is released as carbon dioxide gas. The pressure of carbon dioxide and water builds until the breaking point is reached. A fissure opens up in the surface and a tremendous underwater volcanic eruption occurs, spewing molten lava out to create a brand-new island archipelago, and also releasing hot gas, including the carbon dioxide carrying our carbon atom, out into the evolving atmosphere.

Perhaps a billion years has now passed since the Big Bang created the protons and neutrons that now make up our carbon atom. This atom has participated in the birth of four different stars, and been a part of two of them. It has been processed from individual protons and neutrons into helium, and now into carbon. It has been part of a multitude of dust grains, and part of a blazing comet. It has smashed into a new planet and experienced the force of a million hydrogen bombs. For longer than humans have thus far walked on Earth, our carbon atom was trapped deep underground in its new home. It is has now re-emerged as part of the changing atmosphere of a planet where it could remain until almost the present day. All of the raw materials are here to form the basis of a paradise.

But this is not to be — not this time, anyway. This brave new world is not our Earth. It is instead destined to become a place of lost hopes and dreams, of would-have-beens and should-have-beens. The galaxy will intervene. It is not yet ready for the miracle of life. Within a billion years, outrageous fortune will lay down her fickle hand and eject our atom from this paradise, taking all hope of life in this world with it.

10.
THE GALAXY
STRIKES BACK

There are too many stars in some places, and not enough in others, but that can be remedied presently, no doubt.

MARK TWAIN

In the past several years the news has been filled with heart-breaking stories of genocidal campaigns of terror in various corners of the world, and of the loss of young lives to gun violence in the United States. Whenever I hear such things, the greatest sadness arises when I ponder what might have been. What contributions could these people have made to their families, and to the world, had their lives not been snuffed out before their time?

Yet, as gruesome as humans can be, nature can be far less considerate of the value of life, human or otherwise. A single natural catastrophe can wipe out whole species. But if one mourns for so many lives that are lost, how do we respond to billions of lives that never were? Throughout the history of the universe, and the history of our atom, lost opportunities have been as frequent as the miraculous accidents that have led us to the present moment.

On that lonely planet far, far away and long, long ago, everything was primed for life. But the galaxy had other plans. In fact, the galaxy as we now know it did not yet exist. Recall that our original 9 nuclei were located in a clump of gas some 20,000 light-years across. This spherical gas cloud grew by a factor of 2 or so before beginning its contraction back down to 10,000 to 20,000 light-years across. Then, as stars formed within this structure, and early stellar explosions ejected gas, the whole system began to evolve. Within a few hundred million years, the gas began to collapse toward the central plane of the evolving galaxy, so that the galaxy began to look more disklike. The fourth-generation star that the planet our atom is on orbits around resides in this disk. Yet the entire mass of this conglomeration is less than one quarter of the mass of our present galaxy.

Other clumps of gas have also collapsed, close to the one in which this new solar system is located. In these nascent galaxies, too, stars have formed and died, and atoms have been born and evolved. The separate island universes in which these systems live are precisely that. Just as the Andromeda galaxy, the nearest large galaxy to our own, is a mere smudge in our sky at night, almost indistinguishable from the background stars, had anyone been around on that early young planet to enjoy the view at night the other small galaxies in its cosmic neighborhood could easily have been obscured by dust and local starlight.

But slowly, inexorably, on a timescale so long that any living system would never notice it, all this would change. Over hundreds of millions of years, the small smudge in the sky that might have signaled the neighboring galaxy would have increased in size, growing brighter. Within a billion years, its visible size would have grown sufficiently so that individual stars might be resolved in it. Soon, the stars in this new galaxy would become indistinguishable from the other stars in the night sky. This is because the two galaxies are merging as one. Within another billion years, this clash of the titans will have been completed. The colliding galaxies will have separated, proceeding again on their own. And while there is so much empty space inside each that the actual number of individual stars that collide as the two galaxies cross through each other is minuscule, after they pass through each other the two galaxies will never be the same.

At a distance, gravity still works its wonders. Tidal forces act on individual stars, similar in spirit to the force of the moon on the Earth's oceans, causing high tides. In this case, some stars at the edge of the galaxy are pulled far outside their normal galactic orbits as the neighboring galaxy approaches. A tail of stars begins to follow the trajectory of the new galaxy. As each galaxy continues to rotate, these tails curve, forming spiral arms.

In a similar way, spiral galaxies may die. When galaxies collide head on, the impact is so severe that the two systems can merge and relax into a homogeneous configuration, like a huge cloud of stars. Five billion years from now, the Milky Way and the Andromeda galaxies will collide, and perhaps produce a mishmash that eventually will settle to form a rather featureless elliptical galaxy. But once again, we are getting ahead of ourselves.

Our carbon atom is on a planet orbiting a star at the edge of the disk of the smaller of the two colliding behemoths. As the neighboring galaxy passes through, our star is swept up by gravity into the new emerging spiral arms of this more massive neighbor. The larger galaxy has cannibalized the smaller, and our star has gone along for the ride. When the collision is complete, the galaxy we call the Milky Way is almost fully formed, and our star has a new home. Over the course of the next few billion years, the Milky Way will gobble other smaller satellite systems, as it grows to its present size of more than 100 billion stars.

Surprisingly, it is possible to pluck a star from its orbit in one galaxy and drag it to another without affecting the motion of the planets that orbit the star. During a galactic collision, individual stars almost never come close to colliding. Almost. With a probability of perhaps one in a billion, two individual stars can approach each other closely enough so that the planets surrounding them can be significantly disturbed, or even ejected from their orbits. Our planet is one of the unlucky ones. Within a decade of its ejection from the solar system, it has become a frozen wasteland. Its future is sealed. The great civilizations that might have walked its surface will never be born. Moreover, the tremendous heat it experiences as it swings close to its host star on the last cometlike orbit before ejection is great enough to blow off much of its atmosphere. Our carbon atom finds itself once again without a home, propelled by a solar wind into the vast darkness of space.

Four hydrogen atoms residing in the new host galaxy have over the past billion years had a much less exotic history than our carbon atom. For much of this time, they have remained in diffuse interstellar space, on the outside edge of the rotating gas mass, avoiding the hustle and bustle of star formation, and the concomitant agony of stellar death. But they will not remain immune forever. The stresses induced on the gas as the galaxies collide result in an active new period of star formation. These 4 atoms are swept up with the tide in a huge star, 50 times the mass of our sun. Within a million years the star explodes, and the interior, far too massive to resist the onslaught of gravitational collapse, contracts to form a black hole, from which nothing, not even light, can escape. Fortunately, our 4 particles, now the nucleus of a helium atom, are ejected by the shock wave associated with the supernova explosion, and thus avoid a fate that would remove them from our visible universe, possibly forever.

The stage is now set for the ultimate formation of the atom we find today on Earth. Over the next 3 billion years, our carbon atom and the new helium atom, adrift in the evolving galactic sea of stars, will somehow find each other. During this time, the large-scale dynamic contortions of our galaxy proceed inexorably. The spiral arms, representing waves of high-density gas, move around the galaxy, creating stars in their wake. Supernova explosions expel gas from the surrounding regions, and sometimes from the galaxy itself, while at the same time compressing material elsewhere as their shock fronts propagate, triggering star formation. The galaxy also continues to grow, merging with or cannibalizing small satellite galaxies. A large black hole forms at the center of our galaxy, eating stars and gas voraciously, and causing huge amounts of energy to be released in the process.

All of these grand-scale evolutions are lost on our 2 atoms, however. For them, life consists of two types of processing: Bombarded by radiation from stars, they can combine chemically with other atoms, or they can be expelled from such combinations. Specifically, they can adhere to dust grains that may grow to form rocks, or they can be evaporated from their surface. Ultimately, they may be caught up in huge molecular clouds destined to cool and collapse to form stars. Life seems to consist of constant heating and cooling. Usually the results are different chemical configurations. Only if the atoms get caught up in stars, however, can their elemental identity evolve.

Our atoms are now located in the plane of the disk of the Milky Way galaxy, the plane that houses the spiral arms and the bulk of the stars we now see in the night sky. The gas they have merged with has relaxed and settled into this plane as the galaxy rotates. Amid the large-scale rotation there is plenty of room for individual peculiar motion against or across the flow. On such seemingly random paths, our 2 nuclei will drift closer and closer together over the eons.

Over the course of the first 5 billion years in the life of our galaxy, more than 100 million stars end their lives in supernova explosions. The gas bubbles ejected from each such explosion merge with those expanding from earlier cataclysms. Eventually the products merge with the background gas, and all direct evidence of the life of a star, and its surrounding dependents, is consigned to the dustbin of history.

Yet all is not lost. The existence of atoms, indeed of essentially all the atoms on Earth today, stands as testimony to the many stars that sacrificed themselves so that we might enjoy our moment in the sun. The ultimate merger of carbon and helium to form the oxygen atom that is the focus of our imagination here could have happened at any time in the first 5 billion years of our galaxy's existence, in any of the millions of unstable stellar furnaces that were primed, once they formed, to explode. The abundance of elements such as oxygen and iron has been building up steadily over time. We see old stars on the outskirts of our galaxy that contain a hundred times less iron by weight than our sun does. Only after many, many supernovae could the abundance of heavy elements build up to the fractions we observe today in our region of the galaxy. Nevertheless, my sense of drama makes me suppose that our atom was finally formed in the very last supernova whose products directly created our very own solar system.

It is not too large a stretch of the imagination to suppose this is the case. After all, among the complex atoms on Earth, many were created in the supernova explosion whose expanding bubble came to rest amid the material from which we are now made. We also know that among the products of this supernova the third most abundant element is likely to have been oxygen, followed closely by carbon. In some supernovae, carbon slightly beats out oxygen, as in the explosion that produced the carbon progenitor of our atom adrift in the galaxy. But as oxygen on average beats out carbon in the census of elements now existing in the universe, it is reasonable to assume that this last supernova we will focus on went with the flow, and produced more oxygen than carbon.

Let us thus imagine that the oxygen atom that is the hero of our story achieved its final form just in time. In a cycle that has become familiar, and is carried out in the galaxy with clocklike regularity, our carbon and helium nuclei have found themselves in a dense molecular cloud. They will again be captured, this time together, in the whirlpool that signals the creation of a new star, destined to blaze with nuclear power, and ultimately explode. This time our nuclei will not be quiescent observers, nor will they be captured in the hot neutron star remnant of the burnt-out stellar core. Instead, the carbon and helium nuclei will join together in the last gasp before the star explodes. During the final 10,000 years in the life of this star, a short blip in the 5-billion-year-long saga of these atoms to date, they will collide and fuse. The result will be the nucleus of an atom of oxygen, and a little bit of energy, which will forestall by a moment the ultimate collapse of the host star. Once the race to stellar oblivion is completed, our oxygen atom will be blown out into space on a direct interstellar voyage to Earth, or what will become Earth.

Each atom of oxygen on Earth, by its very existence, suggests a veritable treasure trove of detailed history: the life and death of millions of stars, the slow dynamic evolution of our galaxy, and indeed the history of matter from well before galaxies existed. Our oxygen atom began life as 16 particles. Then, like the Little Indians in the old nursery rhyme, they quickly became 13, as 1 nucleus of helium formed in the first moments of the Big Bang. A few hundred million years later, they were 10, as another helium atom formed. Then some time later they were 7 as a third formed, and then quickly they became 5 as the 3 helium atoms merged to become carbon. In this configuration, 1 carbon and 4 hydrogen nuclei, they persisted for billions of years, witnessing the death of stars and planets, and the breakup of an entire galaxy. Finally, 2 particles, the nucleus of carbon and the nucleus of helium, are brought together from originally disparate parts of the cosmos, with completely different individual histories, to make a single nucleus, the nucleus of oxygen.

This is not the end of the line. It is merely a new beginning. The voyage our oxygen atom takes out of supernova hell will resemble so many its predecessors have taken in the past. And like its predecessors, it will not make the voyage alone. Since there are more oxygen atoms than carbon expelled by this supernova, there are oxygen atoms left over after the bulk of these two elements combine to form carbon monoxide. Our atom instead binds with the most abundant atoms around, hydrogen. Taking two partners, it forms a molecule of H_2O, water. And water, as much as any other form of matter, drives the history that is to follow. Within 17 million years, our oxygen atom, as part of a water molecule, will begin to participate in one of the most amazing sets of transformations that have ever taken place in the cosmos, as far as we know. The physics of oxygen and its supernova partner, carbon, makes possible a remarkable chemistry. Carbon can bond in a hugely diverse set of combinations, with bonds of different types for different purposes. Carbon can be the source of energy, or its beneficiary. Oxygen, however, will occupy a very special role in guiding this process. For as far as we know, only oxygen atoms can combine to form molecules with the ability to power a civilization.

Within 100 million years of the moment this new atom was created, chemistry will make possible geology, and together they will result in a completely new "ology," biology, one that for all we know may never have existed before in the universe. Within 5 billion years, self-aware, self-reproducing entities, composed of atoms of oxygen, hydrogen, and carbon that streamed out of that fateful supernova explosion, will embark on an intellectual voyage of unprecedented magnitude. They will be able to trace their own existence to that precise moment in time, and before that through the eons of cosmic history to the earliest moments of the Big Bang. And today, sometime following the recent end of a man-made millennium, powered by oxygen, and fed by carbon, you and I are continuing the voyage.

11.
FIRE
AND
ICE

His soul swooned slowly as he heard the snow falling faintly through the universe and faintly falling, like the descent of their last end, upon all the living and the dead.

JAMES JOYCE

alling snow is a traditional literary allegory for death, or so I was taught in school. But there are more things in heaven and earth than are dreamt of in the minds of high school English teachers. Five billion years ago a cosmic snow fell faintly through our region of the universe, spelling life, not death. Without it, the complex symbiosis that created the conditions by which life evolved on our planet could not have begun.

Our closest neighbor stands out as a stark example of how utterly barren an also-ran can become. We believe that Mars came tantalizingly close to becoming hospitable, but failed. The tragedy is recorded on the planet's surface in scars from ancient rivers. Why do we flourish, while the surface of Mars is a wasteland? Probably because our sister planet was just a little too small. At the same time, the success of life, and geology, on Earth has erased much of its ancient past. Mars, by its failure to remain vital, suffers no such loss.

In December 1999, a group of very dejected scientists and engineers in California were packing their bags and leaving a command post they had hoped to occupy for several years. The Mars Polar Lander had just been lost, as was its companion craft three months earlier. Did the Lander crash onto the frozen surface of the Martian pole? Did it land safely, and then merely topple over? Or did it land safely and begin its post-landing sequence of activities all alone, waiting for a confirmation signal from Earth that it never received? Subsequent analyses suggested the first possibility. Perhaps decades from now, astronaut explorers will venture to the Red Planet and make their way to the south and recover the Lander to piece together its last moments, in a kind of belated FAA crash-site investigation.

At about $200 million per mission, it costs about as much to send a lander to Mars as it does to make a movie about sending a lander to Mars, or about four times more than a Picasso painting fetches on the open market today. These comparisons are not arbitrary or capricious. What art or films at their best do for us is to cause us to reconsider our own place in the universe. This is precisely what, in spirit, the Mars Lander was designed to do. By searching for water on Mars, and exploring for such things as the fraction of heavy water contained in any water discovered, we could in principle obtain vital information about our own origins.

Up to this point in our saga, I have described events primarily hypothetically. I have talked about possible stars, possible encounters, possible planets, and so on. But now we are coming closer to home, when the possible becomes visible.

We rejoin our oxygen atom almost 5 billion years ago as it is speeding along with a hot nebula of gas that is carrying the remnants of an exploding star, perhaps 15 times the mass of our sun, at a speed in excess of 1,000 kilometers per second through the cosmos. This shock wave blasts into the surrounding gas, losing energy all the while. As the gas temperature decreases below about 1,300 degrees Celsius, the heavy elements iron and silicon, along with carbon, begin to condense into microscopic dust grains. This occurs within a few years of the supernova explosion itself. Eventually this dust shield will obscure most of the light produced by the glowing supernova core. As the shock wave blasts into the surrounding medium, it provides energy to further process pre-existing grains, squeezing some carbon dust into diamonds, for example.

Within 100,000 years, the energy of the shock wave has been spent, and the rich supernova brew of ingredients merges with the background interstellar material. By this time the shell may have expanded more than 100 light-years away from the original explosion. At this distance the leftover neutron star will be invisible. But this new cosmic journey is still just beginning, as the interstellar gas clouds are themselves carried around the galaxy. In the interstellar medium on its voyage outward, our oxygen atom was bombarded by radiation that repeatedly ionized it, and broke apart any significant molecular structures it might otherwise have formed. As the shock wave slammed into the surrounding gas, however, our atom fell by the wayside somewhere during the journey. The energy of the wave, transmitted to the gas, caused it to compress, again shielding our atom from the outside radiation. As the inside of the cloud cools, chemical binding begins. Mantles of frozen carbon dioxide and water begin to coat the aluminum, iron, or silicate dust grains, forming tiny snowballs in space.

Our oxygen atom resides on the surface of one such aluminum-oxide-cored snowball. At the surface of these icy grains, the density is large enough so that, powered by radiation energy making its way through the cloud, carbon can react with oxygen, nitrogen, and hydrogen to build organic molecules. Over time, many of these larger molecules will themselves be dissociated by ambient radiation, but nevertheless a powerful brew of raw materials is forming, even as the molecular cloud continues to slowly lose energy and contract.

This is not any old molecular cloud, however. While these processes have happened before, in a different guise, we now care more about the details, because this is *our* molecular cloud! This would not have been obvious at the time, however. No hint yet existed of the star that would one day blaze inside. In addition, it might then have been located on the far side of the galaxy from where the sun now resides. Over the past 5 billion years our solar nebula has traveled millions of light-years in its voyage around the galaxy. Other stars — including the supernova that gave birth to many of our atoms — that might have once been close to us may now be located thousands of light-years away.

While we might not know precisely where our budding solar system was then, we do know that our planet began to form less than 10 million years or so after a nearby supernova expelled a shock wave whose ingredients now pollute our solar system and may have triggered its formation. Messengers from the past have kindly landed on Earth, providing us with the necessary evidence, frozen in time, of the conditions surrounding the formation of the sun, Earth, and solar system. I again do not refer to aliens, friendly or otherwise, but rather to rocks.

The year 1969 rounded out a decade that was turbulent socially, politically, and scientifically. The world was in the throes of what seemed like a social revolution on an unprecedented scale. Young people were experimenting with drugs, sex, and politics. The United States was pursuing what was turning out to be a very unpopular war. And in the midst of it all, in July of that year two human beings landed for the first time on a solid body outside the Earth.

That same year, the heavens rained down objects around the Earth that would help advance our ideas about the origin of our solar system, and perhaps the origin of life itself. On opposite sides of the Earth, in Allende, Mexico, and Victoria, Australia, meteoritic material weighing several tons was observed to fall to Earth and was recovered. Fewer than one in a million meteorites are recovered after they are seen to fall. Also in 1969, a Japanese expedition in Antarctica discovered that the icy surface was full of meteorites whose abundance had built up over time, perhaps millions of years. This set the stage for meteorite discovery on a vast new scale. Now during each Antarctic summer, as I have described, hardy meteorite hunters scour the high plains near the South Pole for rocks that stand out on the ice like penguins on a Florida beach.

The 1969 meteorites yielded a great deal of information about the early solar system. For more than 150 years, the extraterrestrial nature of meteorites had been accepted. One specific class of meteorites was of particular interest, as it contained substantial amounts of organic materials. In the early 1800s it was commonly believed that perhaps life had been carried to Earth by these extraterrestrial objects, a manifestation of an ancient theory called *panspermia*. In particular, in the first part of the

nineteenth century it was fashionable to believe that life could spontaneously generate, given the right organic conditions, and meteors and
comets seemed to be prime candidates as carriers of such seeds of life.

By the 1850s, rapid spontaneous generation of life was ruled out,
and it became clear that complex organic materials can result from standard chemical reactions in the absence of biology. Interest in these high-
carbon-bearing meteorites then subsided. But the new discoveries in
1969 re-energized the field, by allowing pristine and diverse samples of
material to be tested to determine their age and origin.

The meteorites, named Allende and Murchison, after their discovery
locations, are examples of the high-carbon-content meteorites I described earlier, called carbonaceous chondrites. The adjective is easily
understood, as implying "full of carbon." The noun also has an easily
described origin. These meteorites contain millimeter-size spherical
blobs of rock, called *chondrules*, from the word for "grain," embedded
within them. These small blobs of material solidified as liquid drops
from the gas, at temperatures of 1,300 to 1,600 degrees Celsius. In order
for these to form, the heating and cooling of these systems must have
been very rapid.

In the chondrites, carbonaceous or ortherwise, the chondrules are
surrounded by many individual dust grains and the whole mass is congealed together. The dust grains are of all different sorts and, moreover,
contain material, such as miniature diamond crystals, that clearly
formed in different regions of interstellar space, given the varying abundance of certain rare isotopes of otherwise familiar atoms in them. (Recall that an element is determined by the number of protons in the
nucleus, and nuclei with varying numbers of neutrons are different isotopes of the same element.) Thus material coming from our supernova
progenitor must have scooped up previously solidified dust on its way
here. This dust, in turn — even dust coalesced into a single meteorite — came from many different stars.

Nevertheless, the chondrites have one characteristic, also mentioned
earlier, that makes it clear they are among the oldest, least processed objects in our solar system. The overall abundances of those elements that
are nonvolatile, that is, those elements that solidified at high enough
temperatures so they decoupled from the background gas early on, are

precisely the same as the elemental abundance in the sun. For larger objects, such as planets, a great deal of processing occurs, separating out and enhancing certain elements due to chemical and geophysical activity. Chondrites, however, clearly coalesced almost directly out of whatever gas was available in their region of the early solar nebula.

Which is how we know that at least one supernova very recently predated the formation of our solar system. In the Allende chondrite meteorite, for example, certain elemental abundances such as one isotope of the noble gas xenon, are anomalously large. In particular, xenon-129 is overabundant compared to the present average amount in the solar system. This isotope is formed by the radioactive decay of iodine-129. But iodine-129 decays on a time scale of about 16 million years. This means the meteorite must have coalesced within 16 million years of the time the radioactive iodine was produced. The only site where such heavy elements are produced is in supernovae, and thus we have evidence that at least meteorites were coalescing in our solar system within about 16 million years of a prior supernova. This result is confirmed by searching for excess amounts of other products of heavy isotopes with short radioactive half-lives in the meteorite. For example, anomalously large amounts of silver-107, arising from the radioactive decay of palladium-107, with a half-life of 6.5 million years, have been found.

It is important in making these inferences that the materials considered are heavy enough so that they are created only in supernovae. For example, an excess of one isotope of magnesium is found in meteorites, including the Allende meteorite, arising from the decay of aluminum-26, with a half-life of less than a million years. For some time, this was taken to suggest that the supernova precursor to our solar system occurred less than a million years before the solar system formed. More recently, however, excess magnesium has been observed throughout the galaxy, suggesting that aluminum-26 is continually being produced by sources other than supernovae, perhaps in less exotic and more frequent stellar bursts called novae. Of course, if this is the case, it demonstrates that at least some of the material in the meteorite had its origins in earlier but not too distant cataclysmic events elsewhere in the galaxy.

This is another sign that the ejected material from a supernova combined with a diverse body of interstellar dust already present in space, and this combination, containing our newly minted oxygen atom, began to collapse into what would become our solar system. It is also worth noting that this pre-solar nebula was not particularly well mixed even after the stirring provided by a possible supernova shock. For example, in various carbonaceous chondrite meteorites the ratio of two different isotopes of oxygen found in certain white "inclusions" made from compounds such as aluminum oxide is very different from the oxygen isotope ratios in the surrounding material within the same meteorites. This suggests that these inclusions formed in an isolated oxygen-rich region of the pre-solar nebula that never mixed in well with the rest of the gas, or with the incoming supernova ejecta.

All of this variation on the scale of even our tiny solar system makes it impossible to infer a uniform history for the material we see about us, including the atoms in our bodies. If atoms in a single meteorite arose from different stars, the same may be true for the atoms in the breath you are now taking. We are following an oxygen atom that emerged from a supernova that immediately predated (by 10 million years or so) the formation of our solar system. Other atoms, however, including other oxygen atoms, had very different histories and yet arrived at the same place and the same time. Even once in the pre-solar nebula, the fate of all oxygen atoms is not unique.

The time is now precisely 4.56 billion years ago. Recall that our oxygen atom is now contained in a molecule of water–ice adhering to the surface of a grain of aluminum oxide, which is also perhaps lightly sprinkled with carbon and silicate granules, and topped off with some "dry ice," frozen carbon dioxide.

As the molecular cloud that would form our solar system began to condense, the future of this grain would depend crucially on where it was located in the collapsing pre-solar nebula. We have already followed the general features of the formation of numerous solar systems in which our oxygen atom, in other guises, has played a role. Recall that as gas near the center of the cloud collapses and loses energy, it heats up the medium around the emerging protostar. At the same time, the angular momentum of the collapsing material is removed by polar outflows of gas and dust, and by the complex interactions of magnetic fields and charged particles. Slowly, as the dust and gas collapse inward, they fall toward a central plane of rotating material which will fragment to form planetesimals, and eventually planets that will orbit our sun. The Hubble space telescope has recently allowed us to strikingly confirm this generic picture of star and solar system formation. It has provided, for the first time, high-resolution photographs of the regions around bright, young stars. Clearly visible are bright, circumstellar disks with precisely the sizes and shapes predicted by theory.

But as our atom participates in the formation of yet another solar system, we are naturally much more interested in the intimate details of this particular collapse of dust and gas, because this time around we are the direct by-product of the process. All the atoms in our bodies now were there then. And as we examine things more closely, at least one big mystery arises. In fact, the mystery involves the very material our oxygen atom is now a part of: water. How did the water that now blankets our Earth along with the atmosphere that surrounds it actually get here in the first, or perhaps second, place?

The problem can be simply framed, if not simply answered. As the sun formed, dust and gas collapsed near the center of the nebula over the course of about 100,000 years. The sun began to evolve through a T-Tauri stage of contraction, with huge luminosities and massive stellar winds that would over the course of the next 10 million years or so blow away most of the gas and fine grains located in the solar nebula. Moreover, the temperature of the material as the disk settled into orbit around the sun can be estimated to have been near 1,700 degrees Celsius at the inner edge of the disk, down to temperatures less than the freezing point of water at distances comparable to the present distance of Jupiter from the sun. Near the present location of the Earth, the temperature of the dust and gas was in the neighborhood of 300 to 700 degrees Celsius at the time dust began to coalesce into rocks, and eventually planets. Once the disk had formed, the time frame for large rocks and small planetesimals to form was less than a few thousand years.

The temperature at which material condenses out of the gas, first to liquid, and then to solid dust and rock, is of great importance. Before such condensation, the material is in equilibrium with the gas. Afterward, it accretes into larger grains, its composition is frozen in, and its subsequent evolution is no longer dependent on the temperature of the surrounding gas. Thus the ambient temperature at the time of disk formation will determine the nature of the material that condenses out at each position. Because the Earth has evolved substantially due to many geological processes, one cannot infer directly the temperature of the material that condensed to form the Earth by looking at its present structure. Meteorites striking the Earth, such as the carbonaceous chondrites, provide a much better probe of these early conditions. In particular, a temperature of around 400 to 450 degrees Kelvin (Kelvin, the standard temperature scale used by scientists, is equivalent to Celsius, but with a different, shifted zero point: absolute zero, which is −273 degrees Celsius, is defined to be 0 degrees Kelvin) can be inferred as a temperature limit where carbon compounds are likely to be able to begin to condense, and where liquid water might eventually form, allowing the further buildup of organic compounds within the system. By examining the range of distances of the asteroid belt, from which many of the present-day meteorites emerge, and comparing the distance of origin of most carbonaceous chondrite meteorites versus the distance of origin of their lighter, ordinary chondrite cousins, one can infer an approximate distance of about 2.5 times the present orbit of the Earth as marking the point where the formation temperature of meteorite rocks fell below about 400 to 450 degrees Kelvin. This in turn suggests a rock-formation temperature in the region of the Earth's orbit today of at least 600 degrees Kelvin, and perhaps closer to 1,000 degrees Kelvin. At such temperatures, water, and other volatiles like carbon dioxide and nitrogen, would be effectively baked out of the primordial grains, and would exist as vapor and gas.

But if this is the case, then the rocks that coalesced to form the Earth were essentially devoid of water and other volatile gases. So where did the water needed to make the oceans, and carbon and nitrogen needed for our atmosphere, come from?

Considering the temperature where water–ice could remain congealed around cold dust grains provides one clue. For temperatures lower than about 250 degrees Kelvin (close to the freezing point of water), ice can continue to accrete around grains, which can quickly build up to snowballs, and to the embryo of a planet. The temperatures inferred above for the present region of the Earth and asteroid belt suggest that roughly in the present position of Jupiter a large icy planet should quickly begin to form. Once an embryo greater than about 10 times the mass of the Earth forms, it can rapidly capture a thick atmosphere of gas around it. Thus the fact that Jupiter contains large amounts of hydrogen gas is quite important. Within about 30 million years, the T-Tauri neonatal sun will have dispersed all of the nebular gas in the solar system, particularly the light hydrogen gas. Hence Jupiter can contain a significant hydrogen atmosphere only if it formed prior to the expulsion of hydrogen from the nebular gas around the emerging sun.

The rapid growth of a large gas planet such as Jupiter quickly depletes all of the dust and gas in its region of the solar system, as this material accretes onto the growing planet. As we have seen in another, now long dead, solar system, however, it has a far more interesting effect on the material with orbits just beyond the range where direct collisions with Jupiter are likely to occur. The gravitational influence of the growing planet will serve to perturb the other icy rocks and planetesimals in this region, causing many of them to be ejected from the solar system, and others to be kicked to orbits well outside the solar system. These icy objects make up what we observe today as comets. Every now and then, one of these objects will be perturbed from its large-radius orbit, and will briefly pass through the inner solar system, being observed, if there are observers, from the inner planets as a brilliant object in the sky, with a long tail of volatile gases emanating from it.

As Jupiter kicks the icy comets on their way, many of them will pass through the inner solar system and collide with any planets forming inside the radius of Jupiter. Bombardment of the inner planets by comets, full of water and carbon dioxide, provides one way of transporting water to systems that may initially have been quite dry.

Some calculations of the amount of water, carbon dioxide, and nitrogen that it is possible in principle to transport by early bombardment of billions of comets to a planet in the region of the location of the Earth suggest that these could in principle deliver as much as 10 to 100 times the amount of these materials as presently exists on this planet. If these estimates are correct, then even if the actual delivery mechanism were inefficient, our oceans and our atmosphere could have been created by primordial bombardment of the nascent Earth by huge snowball-like comets.

There is a problem, however, which suggests that at least a significant fraction of our atmosphere and oceans had to have been stored as part of the primordial embryonic Earth. If all the water observed on Earth came from comets, the composition of the water should be similar to that observed in present-day comets. Recent astronomical observations over the past five years of comets such as Halley, Hyakutake, and Hale-Bopp put a wrench in the works, however. If one measures the ratio of deuterium, the heavier stable isotope of hydrogen, to normal hydrogen in water (namely, the fraction of naturally occurring heavy water in the oceans), one finds that this ratio is around twice as high in cometary water as it is in water on Earth.

Is this a problem? Well, there are large uncertainties in the existing measurements, but they are at the very least suggestive. Moreover, one can estimate that the ratio of deuterium to hydrogen in water vapor in the solar nebula should be a sensitive function of position. If the nebular gas lasted up to 2 million years in the neighborhood of the Earth, estimates have been made that the deuterium-to-hydrogen (D/H) ratio would be about half of the observed ratio on Earth today. This suggests that if somehow the Earth could have captured water vapor that might have existed in its neighborhood as it was forming, before the solar wind expelled it from the inner solar system, the predicted D/H ratio today would be too small compared to the observed value. The solution is clear. If the Earth captured a combination of material delivered by comets — having a D/H ratio that is too large — with material captured from the inner nebula — having a D/H ratio that is too small — then the resulting primordial soup would have been, as it was for Goldilocks, just right.

How might we know whether this theory is correct in the absence of good models that allow us to understand precisely how the Earth might have captured primordial inner-nebula water vapor during its formation? This brings us back full circle to the tragedy of the Mars Lander.

Mars has far less water, and far less atmosphere, than Earth. Being smaller than Earth, it was less efficient at capturing any such material, and moreover, any significant early bombardment by asteroids or large meteors could have driven much of the primordial water from the surface. Estimates of the D/H ratio in Martian meteorites yield a surface ratio comparable to that in cometary material, suggesting that all the remaining surface water on Mars could have originally been delivered by comets. On Mars, however, it is estimated that there has been much less contact between water on the surface and water inside the mantle of the planet than there has been on Earth. Again this conclusion comes from estimates of isotopic abundance, in this case of the isotope tungsten-182, which is produced by the radioactive decay of hafnium-182. The latter material has an affinity for rocks in the crust of the planet rather than the core. The fact that a large abundance of remnant tungsten-182 is found in Martian meteorite samples suggests that it was not diluted by significant geological activity after planet formation — the dredging up of material from the mantle to the surface — as has occurred on Earth.

Thus if pristine samples of Martian rock containing material from its mantle can be measured directly, and the D/H ratio is measured, it is possible that evidence for original, low-D/H, inner-nebula water might be found. The only way to be assured of having uncontaminated Martian rock samples is to go to Mars and get them. So if we are ever to truly understand how the oceans that have nurtured and sustained life on Earth first arose, if we are to understand our own origins on this planet, we may need to ultimately send probes to Mars. It seems to me that the possible payoff is worth the price of a few Picassos.

While uncertainties about the details of the formation of the early solar system remain, the general framework for understanding the chronology of the formation of the Earth is now reasonably well established. And there is little doubt that at least some of the present Earth's crust, oceans, atmosphere, and organic material, and perhaps even several of the seeds of life, were delivered airmail by bombardments during its early history. With this in mind, we can now follow the final voyage of our atom to Earth. Nevertheless, I cannot help first deferring to the wisdom of Victor Hugo, who wrote, in an illustrious piece of historical fiction: "We do not claim that the portrait we are making is the whole truth, only that it is a resemblance."

The bitter cold of space at the edge of the collapsing solar nebula in which our oxygen atom is located, frozen as ice on the surface of a microscopic dust grain, masks turbulent dynamics beginning deep in the center of the collapsing sphere of dust and gas. Here, the convulsing stellar core is shooting out jets of matter and huge quantities of radiation. A primordial tension is being established between dust and gases that will govern the system for millions of years. As the evolving sun starts to shoot out a solar wind, and its luminosity begins to grow by leaps and bounds, gas in the surrounding nebula experiences an outward pressure in addition to the inward pull of gravity. These outward forces are great enough, during the T-Tauri stage of the sun's evolution, to expel most of the remaining gas from the solar system.

The material that remains has condensed into dust grains whose makeup depends on their location. Close to the hot core, compounds such as calcium and aluminum oxide, which have a very high melting point, in excess of 1,500 degrees Kelvin, can condense out of the gas. Further out, when the temperature falls below about 1,400 degrees Kelvin, iron and nickel alloys can solidify. Further out still, as temperatures fall below 1,000 degrees Kelvin, various silicates form. Finally, as the temperatures fall below 450 degrees Kelvin at the edge of the inner solar system, water vapor gets incorporated into the mineral lattice of the grains. Once one goes out beyond about 4 times the present distance of the Earth from the sun, water–ice can begin to coat the grain cores.

Once dust grains grow to a certain size, they are impervious to the pressure produced by collisions of individual gas particles, so that this pressure cannot operate to hold them up against gravity. Dust then begins to "fall" inward.

While some of this material actually falls into the emerging star, much of the dust settles into the swirling central disk that inevitably seems to accompany star formation. Material thus rains down both on the star and on the disk from above and below. In this way, even at points located as close to the sun as the present Earth is located, some dust that has condensed from much farther away will settle downward onto the disk, adding new ingredients to the mix of material that will eventually form the inner planets.

As the dust sweeps past the gas on its infall, it gets a small drag force opposing its motion. This causes particles to spiral further inward. On these new trajectories they will collide with other dust particles. These collisions, if they are sufficiently energetic, can serve to break up any emerging masses, but more often than not, grains coalesce. Larger grains begin to eat smaller grains, and quickly larger objects form. Even before many of the grains have settled down to the central disk, they have grown to become centimeter-size, so that their inner materials are completely isolated from the surroundings, and materials that condensed out earlier remain safely tucked away.

The rich get richer, and the poor get poorer. So it is in the heavens as it is on Earth. Larger and larger rocks form from collisions, and within 10,000 years kilometer-size objects have accreted. The collisions are now quite dramatic, and often generate enough heat to melt the material in the protagonists as they merge. In this manner, the inner planets begin to grow.

Further out, at the point where water–ice can form, Jupiter is growing at an exponential rate. Quickly, within 10 million years, before the stellar wind can eject all the hydrogen gas from the inner solar system, Jupiter has gobbled up more than 100 times the mass of the Earth in rock and ice, and has generated a gravitational field strong enough to capture 3 times that much hydrogen as an atmosphere. Very quickly this giant devours any material in its path, emptying this region of the solar nebula of extraneous material.

Further out still, our oxygen atom has become incorporated into a large rocky snowball of material, well on the way to forming the nucleus of a new planet. Jupiter, however, has beaten all its competitors, and in a relatively close encounter, our snowball swings by, and like a pebble in a slingshot, gets shot outward. A subsequent collision with another ice boulder is so energetic that the two objects break apart into myriad pieces, melting much of the water, which subsequently refreezes. During this continual melting and refreezing stage, organic materials that have already formed as interstellar hitchhikers on the surface of primordial dust grains get incorporated into the body of what will become comets, and new chemical reactions occur in the melted material, building up yet more complex substances. In the cores of these giant dirty icicles, aluminum-26, the radioactive substance discussed earlier, decays on a timescale of less than a million years, providing heat to keep some of the water in a liquid state, mediating continued organic synthesis.

Some of the planetesimals ejected by Jupiter travel through the inner solar system before passing by the sun and heading out again. The great speed with which they head away from Jupiter causes them to impart energy and stir up the materials that might otherwise form planets nearby. Between Jupiter and the small emerging planet Mars lies a vast reservoir of failed planets, whose collisions were so energetic that rather than build up into a large mass, they broke apart into smaller objects tens of kilometers across. There they remain today as a large belt of asteroids, which will, with certainty, come back to haunt us.

Further out, on the far side of Jupiter, the water molecule containing our oxygen atom has remained relatively unscathed, other than melting and refreezing for perhaps 40 million years as its various parent bodies have tumbled and crashed into each other, following close encounters with the giant Jupiter. Finally one last gravitational kick, and it is sent out with high velocity on a trajectory that could carry it 1,000 times further from the sun than Jupiter orbits. There it could become part of what is now known as the Oort cloud of comets, orbiting far outside our solar system, well on the way between the sun and the next nearest star. This region, containing literally trillions of comets today, was largely populated by gravitational kicks from Jupiter and the other giant gas planets as they stirred up the primordial solar system. Periodically the motion of nearby stars or other disturbances knock a comet off its course in the Oort cloud, sending it on a trajectory that will bring it toward the sun to temporarily brighten the night sky. Most comets in this realm will, however, survive for billions of years cast out in frozen limbo at the edges of interplanetary space. Before it could reach the Oort cloud, however, random chance intervened to make our atom's future far more interesting.

During the tens of millions of years our atom spent in a cosmic snow-ball fight near Jupiter, things were heating up considerably in the inner solar system. Here violent collisions immediately signaled the early wars of dominance, as larger bodies swept up and cannibalized smaller ones. As noted previously, these collisions frequently had sufficient force to liquefy the participants themselves. Perhaps in this process molten rock was spewed into space, subsequently cooling quickly to form the chondrules that are found in meteorites, which would once have had to have been free-floating liquid drops. Simulations show that these must have cooled rapidly, and could therefore not have gently melted and then cooled as the background nebula temperature fell.

In any case, after several thousand years of such give-and-take, much of the mass was locked up into kilometer-size objects. From here collisions would begin to slow down, occurring about every 1,000 years or so. Nevertheless, larger objects began to experience runaway growth so that within 20,000 to 100,000 years, many objects the size of our moon existed in the space occupied today by the inner planets.

As more mass got tied up into fewer larger and larger bodies, the rate of collision slowed. Estimates suggest the growth from moon-size planetesimals to Earth-, Mars-, and Venus-size planets would have taken at least 10 million years. Even then, the planets would not have been fully formed. If the final growth stage of the Earth occurred by a short burst of runaway growth, and was assisted in the beginning by a continued presence of nebular gas, it could have been 99 percent fully formed within 40 million years or so.

The evolving Earth certainly lacked much of an atmosphere. In the first place, the high temperatures involved during much of the condensation of rock in the inner nebula would have "degassed" much of this material. Next, during the 10 million to 40 million years of formation, the sun's turbulent wind would have blown off any gas in the inner solar system. This effect, combined with the constant gravitational disruptions caused by collisions and near collisions of massive planetesimals, would have stripped off any nascent gas that might have originally gravitationally settled around the growing Earth.

The biggest disruption of all would have occurred relatively late in this process but still within 100 million years of the Earth's original formation. At this time, the largest planetesimal collision in the history of the Earth is likely to have occurred. An object perhaps the size of Mars grazed the cooling surface and mantle of our planet, ejecting billions of tons of material into orbit around it. The heat generated would have remelted the entire body (if any of it was solid at the time). The material spewed out into space formed, within a period perhaps as short as a few years, what is now the moon. This was much closer to the Earth than it is now, perhaps 3 or 4 times as close, orbiting once every 5 days or so.

This collision can explain many things, including the high rate of spin of the Earth (the grazing collision would have twisted it around like two football players who collide while running in opposite directions). It would also explain why the moon's orbit is inclined relative to the disk of the solar system: if it had collapsed along with the Earth, it would be expected to orbit along the same plane. Finally, it would explain why the moon contains much the same kind of material as the Earth's mantle, and is also largely degassed (the collision would have ripped out mantle material, but volatiles would have been spewed into space by the great heat of the collision).

This collision, if nothing else, would have assured that any pre-existing atmosphere was ripped away from the growing planet. It would also have continued the process of heat generation that would have left our planet bubbling and boiling for much of its first 100 million years of existence. The intense heat from collisions would have been compounded by heat generated by the gravitational collapse of the heavy elements like iron to the core of the Earth, and the heat generated by radioactive decays inside the Earth. The heat generated by all these processes would have been sufficient to melt even the crust of the Earth during these first tens of millions of years.

This heat would have allowed chemical and physical processing of the material that made up the Earth. This process, called *fractionation*, would have left the ultimate Earth looking very different in form from the individual meteorites and planetesimals that made it up. For example, in the liquid state, iron, iron sulfide, and iron oxide would have sunk into the core, as would have nickel and various other elements that tend to follow iron in chemical mixtures.

Another factor, however, one much more important for the ultimate life of our atom of oxygen on Earth, governs the chronology of the formation of the Earth. One would expect that the first material to form the primitive planetesimals that bombarded the growing Earth would have been made up of local matter, condensing out of the gas at high temperature, and thus involving iron and aluminum, and containing very few volatile gases. The core of the Earth therefore could have formed quickly. Later on, as perturbations by the growing outer planets began to stir up the pot, meteorites from the asteroid belt could have begun to pound the growing Earth. These materials contain carbon, and have a larger volatile component. Finally, one might imagine that material from even farther out would have begun to bombard the Earth. This material, cometary in nature, could have contained significant water, carbon, and even organic materials.

There are a number of pieces of evidence that support the idea of a quick formation of the core, and subsequent rapid buildup of the outer portions of the Earth. First, in the outer mantle one finds a larger than expected composition of elements like nickel and cobalt that tend to follow iron when in molten conditions. The fact that these elements remained outside the core suggests they may have been deposited there after much of the core formed.

Another bit of evidence that the core formed relatively quickly involves measurements of the noble-gas composition of the atmosphere of the Earth. The molten material of the Earth would have released gases out to the growing terrestrial atmosphere. If the development of the core had been slow, these gases would have built up over a significant time period. Noble gases are a very good tracer of such early "outgassing" of the Earth because they do not react chemically, and thus once released into the atmosphere, they stay there (except, of course, for the lighter elements that float to the top of the atmosphere and are lost). The noble gas argon comes in primarily two different stable isotopes, argon-36 and argon-40. The latter occurs only via the radioactive decay of potassium-40. If the core built up slowly and the covering mantle remained molten along with it for a long period, then the potassium-40, which has a half-life of about 1 billion years, would have had time to decay substantially. It would have then been able to release significant amounts of argon-40 to the atmosphere. But the ratio of argon 40 to argon-36 in the Earth's atmosphere is almost 100 times smaller than that in material trapped in the present Earth's mantle, indicating that most of the gas that was released by the early molten Earth was released within a period of perhaps 10 million to 100 million years.

If the iron core of the Earth formed relatively rapidly, then the subsequent buildup of the Earth, especially the outer layers, ultimately including the atmosphere, could have been substantially affected by the later delivery of material from meteors and comets. We can calculate that Jupiter would have taken several hundred million years to eject most of the proto-cometary material in its vicinity out to the Oort clouds or completely out of the solar system. During this time, of course, it would be peppering the inner solar system with material as well. Neptune and Uranus would have ejected material with a somewhat longer timescale. Thus we would have expected that for at least 100 million years, continuing at a reduced rate for up to 500 million years, an unending bombardment of material would have stirred up the surface layers of the inner planets and led to at least partial melting of the crust. Observations of the moon's surface, which has preserved the early cratering, unlike the more dynamic Earth's surface, provide ample evidence that the first few hundred million years of history on Earth were ones of constant catastrophe. Objects up to several hundreds of kilometers in size were raining down on the planet perhaps as often as once every few million years.

As the Earth cooled, any gases trapped inside the molten material were being released, and from the deuterium data in ocean water we know that a significant fraction of the present water now on Earth was primordial, contained in the original planetesimals that collided to form the Earth. We also know that the material that bombarded the Earth in its latter stages of formation must have delivered a somewhat lesser fraction of the water, but still a significant component.

These impacts were not benign. A 300-kilometer-diameter object colliding with the Earth, for example, would provide enough heat to evaporate all of the world's present oceans, and heat the entire surface of the Earth to a temperature in excess of 1,300 degrees Kelvin. It might take more than 1,000 years for this water vapor to cool and condense back into liquid form.

And so it is that our oxygen atom finally makes its way to Earth, experiencing déjà vu. Once before, 1 billion years earlier, on the other side of the galaxy, it took such a ride. Now, kicked out from its orbit near Jupiter, the ice-covered rock containing our oxygen/water molecule starts its journey through the solar system. By an accident of fate, it moves through the inner solar system on its voyage past the sun and out to the outer regions of the growing Oort cloud. But it never makes it past the sun. The icy comet, with a tail of material outgassing behind it, finds the Earth amid the emptiness, like a needle in a haystack. Within 10 years of starting its new journey, our atom finds itself in an object on a collision course with what is destined to become a blue planet. Much of the outer shell of the comet has been shed by the time the bulky remainder hit the top of the growing Earth atmosphere. Material in fine grains that have been so shed can rain down relatively slowly through the atmosphere, without generating much heat. In so doing, it is quite possible that some of the organic material created in interstellar space, supplemented by that cooked in the cometary kitchen, might make it down to the growing ocean unscathed.

Our oxygen atom, however, is buried deep inside the comet. The intense heat generated by the comet's voyage of destruction through the atmosphere breaks it apart into smaller pieces, but these survive the voyage all the way to the ground. Some crash into the newly formed crust, locally melting it and piercing well down into the surface material. The last time our atom collided with a planet, it hit dry land. This time the piece containing the water molecule housing our oxygen atom crashes into the newly forming ocean, boiling off large quantities into steam, carrying our atom in the water vapor dispersed into the atmosphere. All in all, the comet has released into the atmosphere an energy equivalent to more than 1,000 megatons of TNT.

Our atom has now completed a circuitous 5-billion-year voyage to Earth. Here it will stay longer than its constituents have ever stayed in any single location in the history of the universe. At least for the time being, our atom has come home.

12.
COOKING
WITH GAS

*The science of life is a superb and
dazzlingly lighted hall which may be
reached only by passing through a long
and ghastly kitchen.*
CLAUDE BERNARD, AN INTRODUCTION TO THE
STUDY OF EXPERIMENTAL MEDICINE *(1865)*

The excitement was palpable when on March 12, 1610,
Galileo Galilei announced to the rest of the world that
a hidden universe existed just beyond the reach of our eyes. On the day
of publication, the British ambassador to Venice dispatched a copy of
Galileo's new book to King James I, promising him: "The strangest piece
of news (as I may justly call it) that he hath ever yet received from any
part of the world; which is the annexed book (come abroad this very day)
of the Mathematical Professor at Padua, who by the help of an optical instrument . . . hath discovered four new planets rolling about the sphere
of Jupiter, besides many other unknown fixed stars."

I think it is hard for anyone living at the beginning of the twenty-first
century to truly appreciate how remarkable it must have been to suddenly learn that even our own solar system was not what it seemed to be.
Suddenly, four new neighbors of the Earth revealed their existence.
Could there be many more? And if the rest of the solar system "rolled
about" the Earth, why did these four new interlopers orbit Jupiter? Even
the vast power of the Catholic church at the time could not stop the revolution that was about to unwind as a result of this simple observation by

a mathematician (for there were not yet "physicists") at a small but renowned university in Padua.

To try to understand the impact of this revelation at that time, imagine that evidence is unearthed (or more accurately, "unmarsed") today implying that 2 billion years ago intelligent life had flourished on Mars, only to die out without leaving a visible trace on the planet's surface today. The shock would be enormous, and the implications of such a discovery would be likely to shake theological foundations at least as much as those of Copernicus or Galileo ever did.

Of course, everything we know about Mars, and everything we know about the evolution of life, argues against such a possibility. Nevertheless, I once had a debate with a reader of one of my books on the subject of possible past intelligent life on Mars. This individual was no kook. He wasn't claiming, for example, that the afternoon shadows that accidentally produce, from a certain angle, what appears to be a face on the Martian surface was, as others have claimed, evidence of some early lost civilization. Rather, he was arguing just the opposite. We know that billions of years ago, liquid water flowed on a warmer Martian surface, and the planet may have seemed ripe for life to evolve. His point was that no visible trace of any civilization that may have lived and died over 2 billion years ago would be left, due to the ravages of time. Therefore, how could we dismiss this possibility?

On the surface (if you will forgive the pun), this argument cannot be dismissed out of hand. Two billion years is a very long time. However, Mars has far less geological activity than Earth does, so the crust of the planet is not regularly recycled. Thus, while obvious evidence from space of ancient cities might be difficult to detect, evidence for past intelligent life on the surface should be possible to uncover even utilizing unmanned probes. This is not to mention the fact that, on Earth at least, it took more than a few billion years for living systems to evolve to become self-aware.

Nevertheless, this got me thinking about Earth. If we annihilate ourselves tomorrow, will there be any significant evidence 2 billion years from now of our ever having populated the planet?

The Earth is, after all, a dynamic planet. Our present continents are newcomers on the scene. Moving at roughly the rate at which fingernails grow, they are drifting apart in a measurable way. Two hundred million years ago they were joined together in a single supercontinent, now labeled Pangaea, and several hundred million years before that in a previous supercontinent. In the intervening years, the continents pulled apart, then rejoined in different patterns. The material of the continents is floating on a layer of denser rock, and convection forces in the planet are causing the Earth's crust to be recycled over the course of hundreds of millions of years. Material at the interface of colliding continents is driven down, or *subducted*, into the mantle below, to be heated and melted by the intense heat and pressure deep down in the Earth, while fresh material is spewing out of volcanic ridges in the middle of the oceans.

Will there thus be any direct trace, on the planet's surface, of New York City, the Pyramids at Giza, or the Great Wall of China 2 billion years from now? Probably not, although some buried artifacts might survive this turmoil.

This is not to say that a future alien paleobiologist who visited this planet wouldn't be able to deduce that complex living organisms once roamed its surface and perhaps changed its environment over their 4-billion-year heyday, even if no artifacts were discovered. But it is nevertheless not so clear that if humanity were to perish soon, our existence would ever be noted in anyone else's galactic history books.

There may be hope, however. We face the same challenge today if we try to look back billions of years to unravel the mysteries of our emerging planet. But new relic evidence of the early moments of the Earth's history is constantly being uncovered, so that previous large gaps in our understanding are being filled in, and a coherent 4-billion-year history is evolving.

It is fitting that a key geologic message, frozen in time, is found in frozen ground. At the edge of the inland ice at Isua in southwest Greenland lies an innocuous outcropping of rock, calcite quartz, and clay, combined with some iron compounds and small amounts of organic carbon. It is large enough to be noticeable, yet on a grand terrestrial scale it is minuscule, only 40 kilometers long by a few kilometers in width. Nevertheless, buried here are the oldest known sedimentary rocks in the world. Three independent dating techniques all imply that these sediments are almost 3.8 billion years old. More important, the existence of the sediments implies that at that time this was a world with liquid water oceans, covered by an atmosphere containing carbon dioxide.

In a world in which the continents are regularly being recycled, it is wonderful luck to have objects dating back 3.8 billion years to help us sort out our past. But let's be greedy. Three billion eight hundred million years is old, but it is still more than 700 million years *after* the Earth accreted from the solar nebula. What about the intervening time?

Actually, there exist remnants that are even older: small crystals of zircon dating back to the igneous rocks that weathered to form the sedimentary rocks of the type seen at Isua. Individual crystals of zircon found in much younger sediments in the Jack Hills of western Australia date back almost 4.3 billion years, less than 250 million years or so after the formation of the Earth.

But these are not old rocks, just old crystals embedded in younger rocks. The oldest complete geological structures were recently discovered on the banks of the Acasta River, in the barren wilderness of Canada's Northwest Territories. Why old rocks should have such an affinity for cold places is beyond me, but they seem to be found there. A lone geologist's shed stands next to the river, with the welcome sign "Acasta City Hall, Founded 4 Ga" painted above the door (Ga stands for giga-annum: billion years). Recent dating of the igneous samples from the area yields an age just over 4 billion years, 4.055 billion years to be exact.

The mere existence of these rocks on the surface today is significant, because it tells us that some form of early continents was present even at this time. After all, continents do not form until lighter minerals cool and form a crust that floats on top of the molten flow beneath. And the early Earth was hot!

Following the accretion of moon-size boulders, and the rain of comets and asteroids that kept the Earth molten, forming its core and ultimately building up its mantle, the barrage did not stop. Today there are about 1,000 objects in Earth-crossing or nearly Earth-crossing orbits that are larger than about 1 kilometer across. The rate at which 1-kilometer-size or greater objects strike the Earth on average today is about once per million years. The rate at which such impacts occurred after the Earth first formed was almost a million times greater. Thus every decade or so on average, for at least 50 million to 100 million years, a massive object slammed into the Earth! Of course, one of these massive objects was the comet that delivered our oxygen atom to the Earth's atmosphere. But it was far from the last. What was the effect of such bombardment?

A 1-kilometer comet or meteor hitting the Earth first produces a crater about 20 kilometers across. Weathering makes it unrecognizable within 600 million years. A 10-kilometer meteor, by similar reasoning, produces a crater about 200 kilometers across, and this can remain visible for more than 2 billion years. But craters are not all that are produced. The extent of the violence was well described in the year 1178, when five observers of the crescent moon reported the creation of a crater later called Giordano Bruno, in honor of the Italian philosopher burned at the stake, who had proclaimed his belief in the existence of other planets in the universe. The observers wrote: "Suddenly the upper horn split into two. From the midpoint of this division a flaming torch sprang up, spewing out over a considerable distance fire, hot coals and sparks. Meanwhile the body of the Moon which was below writhed as it were in anxiety . . . and throbbed like a wounded snake."

Bombardment by comets and meteors heated up the growing atmosphere of the nascent Earth, and its surface as well. As I described in the last chapter, judging from the cratering of the moon's surface, objects as large as 200 kilometers bombarded it until at least 4 billion years ago. Because the Earth is much larger than the moon, we can estimate that perhaps two dozen objects this size, about one every 10 million years or so, struck the Earth following its creation and up until it was about 400 million years old. Each such impact would have evaporated most of the world's oceans, and heated the crust up to in excess of 1,000 degrees Kelvin, locally melting parts of it. Also, the steam created would remain in the atmosphere for more than 1,000 years. In addition to dumping large quantities of water and other volatiles into the atmosphere, the impact of the collisions could also embed some volatiles from the colliding objects into the Earth's mantle.

Smaller objects would fall with much higher frequency. A 10-kilometer object landing in the ocean would produce worldwide tidal waves. The heat generated by the debris following impact would fill the sky with a glowing incandescence for hundreds, if not thousands, of kilometers.

Chemical reactions would produce new molecules in the atmosphere. Such a collision would be expected to have occurred every 20,000 years or so for perhaps the first 400 million years of the planet's history. As we shall later see, it is just such a collision that is thought to have produced massive extinctions on Earth about 65 million years ago, including wiping out the dinosaurs. Collisions with smaller but still deadly objects would have occurred even more frequently.

During this period, therefore, much of the water above the surface of the Earth would have remained in the atmosphere. There may have been short periods of quiescence, when our atom, in its water molecule, may have condensed on solid ground, but these were few and far between. Early on, much of the surface was probably a magma ocean of molten rock, and parts of it were remelted at regular intervals, even as it cooled. The molten rock would release its volatiles to the atmosphere, building up the carbon dioxide and water levels further. In short, our oxygen atom had entered a steamy version of hell.

The rate of comet and meteor bombardment fell off exponentially with time. The rate of cometary expulsion by Jupiter would have fallen off to 1 percent of its original value within 400 million years, and by 3.5 billion years ago the net impact rate from all sources would have fallen by a further factor of 100. Thus by about 4.1 billion years ago the rate of heating was small enough that the Earth's surface would have essentially solidified, except for random local impact catastrophes. The rocks from northern Canada are survivors from this earliest era of continental formation.

This is not to say that things were not still quite hot, and indeed, this phase of the Earth's formation, up until about 3.9 billion years ago, is appropriately called the *Hadean period*. The core and mantle were probably still sufficiently hot from the heat delivered by impacts and gravitational settling, but also due to the huge radioactivity levels still present, so that any crust that formed on the surface was still thin.

If you fly into Cleveland's airport and drive through the city on the way to my university office, you will drive past the last remains of the region's previously booming steel industry. On some days a huge fire can still be seen burning from atop a large smokestack, looking like a landlocked lighthouse. In the factories, steel is processed in slag furnaces. On the surface of the molten steel a thin flexible skin forms. Any crust on Earth in those early days would have resembled this thin skin.

During this early period, the radioactivity level in the Earth was about 4 times higher than it is today. The convection currents in the mantle were thus significantly stronger than they are today. The large solid plates on which the present continents float, responsible for the global plate tectonics that mold and remold the face of the Earth on regular intervals, had not yet built up.

The radioactivity levels during these early times were higher simply because the dominant radioactive elements, uranium, thorium, and potassium, have half-lives ranging from 700 million to 14 billion years. Thus there were simply more of these radioactive isotopes around at that time. Incidentally, the distribution of radioactive material in the Earth is surprising, or at least it was to me the first time I heard about it. You might think that the heavy elements like uranium would sink down to the core when the Earth was molten. But uranium atoms, when they are ionized, as they would have been in the liquid magma, are very large. This causes them to float toward the surface. For this reason it is currently believed that most of the radioactivity left in the Earth is located in the continental crust.

I should point out that this is a supposition with important consequences. Indeed, one can work out that a thin layer, 10 kilometers in depth, covering the Earth, and with the present average radioactivity of granite rock, would produce enough heat to account for the entire heat flow coming from the Earth at the present time. If there were significantly more radioactivity in the Earth's mantle and core, the Earth would actually thus be heating up at the present time! As far as I know, no one believes this to be the case, but nevertheless, 15 years ago I proposed a set of experiments that could empirically constrain the radioactivity levels inside the Earth. It turns out that the radioactive decays of naturally occurring isotopes in the Earth all lead to the production of antineutrinos, the antiparticles of the weakly interacting particles bombarding the Earth every day from the sun. If one could measure the surface flux of such particles, which can travel right through the Earth following their production, one could infer the abundance and distribution of radioactive elements in the Earth. This is a very difficult set of experiments to perform, however, and to date no detector is yet sensitive enough to probe this potentially useful signal.

By about 3.8 billion years ago, around the time the Isua sediments formed, things were beginning to settle down considerably. Oceanic crust had solidified, and the last ocean-evaporating event had probably occurred, although events that caused the upper few hundred meters of the ocean to evaporate could have continued up to a few hundred million years later.

Were it not for the existence of Jupiter, and to some extent the other large gaseous outer planets, things might perhaps have never settled down on Earth, at least not sufficiently for any of its subsequent history to occur. By kicking out cometary debris within a period of a few hundred million years, Jupiter assured that the rate of impacts on Earth would fall off exponentially with time. Had Jupiter not accomplished the task of creating the present-day Oort cloud, the rate of cometary and meteor bombardment of Earth could have been perhaps 10,000 to 100,000 times larger, throughout its entire history, than it is today. In this case, ocean-evaporating events would have continued at a rate of at least one every 10 million to 100 million years throughout the Earth's history.

Thankfully, however, things did settle down. As the steamy atmosphere began to condense, a hot rain poured down upon the Earth, not for 40 days and 40 nights, but for 40 million days and nights, or even longer. The water molecule containing our oxygen atom would have fallen in this rain, helping create the oceans that began to cover the Earth. But even in the oceans, things were more turbulent then than now. The Earth was still hot, and the energy pumped in to drive ocean currents was intense. In addition, the moon was perhaps 4 times closer to the Earth than it is at present. Orbiting in a period of 5 days or so, it would have produced huge tides, 30 times as large as at the present, and this would have also contributed to subsequent heating of the oceans.

Some water vapor remained in the atmosphere, but the dominant gas at this time was undoubtedly carbon dioxide. Released from the impacting comets and from a molten Earth with heartburn for millions of years, Earth's early atmosphere was perhaps 60 times more massive and quite different in composition than the present atmosphere. As primordial hydrogen would have already evaporated, carbon dioxide probably accounted for 98 percent of the gas in the atmosphere, with a few percent nitrogen, and some water vapor.

And a good thing it was, too. After the sun settled down to its steady-state long-term hydrogen burning, after its turbulent T-Tauri stage, its early luminosity was only about 70 percent as great as it is today. At this level, had the Earth been surrounded with its present atmosphere, the oceans would have eventually frozen over. Yet there is no evidence at all for an early period of glaciation throughout the Earth. In fact, all evidence is that it was much warmer then than now.

I alluded to the reason for this before. It has its roots in the same concern that is prompting the rich industrialized nations to reconsider their policies toward the burning of fossil fuels. Carbon dioxide is a so-called greenhouse gas. It is transparent to visible light from the sun, but absorbs the infrared radiation that would otherwise be emitted into space by the Earth. As such, it causes global warming.

The greenhouse effect at that time was, however, orders of magnitude greater than anything we might expect to occur today. The reason is that the amount of carbon dioxide in the atmosphere 4 billion years ago was at least 10,000 times greater than it is now. The makeup of the Earth's atmosphere at that time, primarily carbon dioxide and nitrogen, with a pressure of perhaps 20 times atmospheric pressure today, was very similar to that of the present atmosphere of Venus. The blistering surface temperature on that planet is over 750 degrees Kelvin (about 880 degrees Fahrenheit).

There but for the grace of liquid water go we. The total carbon dioxide and nitrogen inventories of the Earth and Venus are almost identical, as are their sizes. Had our water molecule landed on Venus and not the Earth, however, it would have had a very different history. Early on, when the sun was 30 percent less luminous, there may have been liquid water on Venus, but as the sun heated up, so did the planet. As the water evaporated into the atmosphere, it contributed to a runaway greenhouse effect. Both water vapor and carbon dioxide have the same impact. The more water vapor in the atmosphere, the hotter the planet would have gotten. But the hotter the planet, the more evaporated water, and so on. Today, Venus is a dry planet. The total water content is about 10,000 times smaller than that of the Earth. It is thought that the water in

Venus's atmosphere was, over time, broken up into separate hydrogen and oxygen gas by radiation from the sun. The hydrogen would then have evaporated from the planet.

Why didn't the Earth have a runaway greenhouse effect, and why is the amount of carbon dioxide in the atmosphere today a small fraction of what it once was? Liquid water, and luck of the draw.

The Earth's location near the sun has allowed for liquid water to remain on the planet for the past 4 billion years or so. The present temperature is such that were the Earth to warm up a bit, the amount of extra water that could be held in the atmosphere would not be enough to trigger a runaway greenhouse effect. If the Earth were 15 percent closer to the sun, this would not be the case (Venus is, alas, 30 percent closer). Similarly, at early times, when the atmosphere was dominated by carbon dioxide, the sun was 30 percent less luminous, and once again the temperature on Earth, while perhaps the equivalent of hot tea, nevertheless kept the bulk of the water in the oceans.

Why, then, as the sun heated up, did the Earth not get hotter? Here we, and the water molecule carrying our oxygen atom, encounter for a second time one of the most remarkable feedback mechanisms in nature. It will govern much of the history of our atom on Earth, and also much of the history of life on Earth. Recall that carbon dioxide can dissolve in water, and the higher the pressure, the more carbon dioxide water can take up. In particular, carbon dioxide will dissolve in rainwater, forming carbonic acid, H_2CO_3.

This is, by the way, the same principle that governs why sodas fizz only after the bottle is opened, and why they taste flat after awhile. When the soda is under high pressure in the bottle, more carbon dioxide can dissolve in the water, forming carbonic acid. This is the substance that gives sodas a tart taste. Once the bottle is opened, the pressure is reduced, and the water can hold less carbon dioxide, so that it bubbles up out of the water, creating the fizz. In turn, the carbonic-acid levels in the water decrease, making the soda taste flat.

Now the acid water will attack silicates in rocks, forming precipitates that yield carbonate rocks such as limestone and dolomite. These will sediment out to the ocean floor, thus effectively removing carbon dioxide from the atmosphere. Moreover, it turns out that carbonate formation is very temperature dependent. The hotter the temperature, the more efficiently carbonates form, and the more effective this mechanism is in removing carbon dioxide.

This mechanism, operating over billions of years, was able to lower the carbon dioxide levels in the atmosphere by a factor of 10,000, as the sun in turn increased in luminosity and kept the planet warm. In this way, nitrogen slowly became the dominant gas in the Earth's atmosphere. Also in this way, sedimentary rocks began to form as the carbonate materials were laid down. The existence of the Isua sediments indicates not only that water oceans existed 3.8 billion years ago, but also that the removal of carbon dioxide from the atmosphere had already begun in earnest.

Our water molecule participated in this process. Part of an early steamy rainstorm following an energetic impact that hurled it into the atmosphere as water vapor, the oxygen in the water was bound to the carbon dioxide, as it had earlier bound to oxygen when it was a carbon atom on another world, forming H_2CO_3. This carbonic acid molecule fell on the newly formed crust of the Earth, eating the silicate away and becoming incorporated into a limestone precipitate that was carried in a rivulet to the rising sea. There, it sank to the ocean bottom, where it would otherwise remain for the rest of its time on Earth, were it not for heat and chemistry.

As the Earth cooled, and the lighter materials floated to the surface by partial melting, a surface crust began to build up and become more rigid. While substantial convection was occurring in the mantle, the crusts that first formed on the surface were too thin and brittle to be pushed around en masse by the dynamics below ground. As these crusts built up, containing mostly very dense basalt, this material formed the crust that lies under the oceans. The Isua sediments formed on top of such crust. Eventually, once the crusts became sufficiently rigid they broke up into large plates that floated on the convecting mantle below, and moved with these convective flows. What has become known as plate tectonics had begun on Earth, and with this new process continents could arise.

In various regions where the mantle convection flows upward, new crust material can be created. Such regions are observed today in the mid-ocean ridges. As this new material forms, the pre-existing crust is pushed apart. In other places, two moving crustal plates collide. In this case, something has to give, and once the plates are thick enough, one plate can literally be pushed under the other, as experienced once before by our atom. This subduction process causes partial melting of the crustal material that is being driven downward. In addition, the volatiles, stored in the carbonates that sedimented down to the crust, are heated, and carbon dioxide and water are released. As the pressure builds up, eventually volcanoes arise, spewing not only these volatiles back into the atmosphere, but also lighter rock that has melted out of the subducted crustal material. This less dense rock, granite, will ride higher on the surface, rising above the oceans, ultimately forming continents.

The mean age of the present continental crust is only about 2 billion years. This does not mean that continents are only this old, as crustal material could have been recycled numerous times. Granites date back more than 3 billion years, suggesting that major continent formation had been carried out in earnest by that time, having begun around the time of the Isua rocks, slightly over 3.8 billion years ago.

This geological cycle of subduction and volcanic activity not only provides a mechanism for forming continents that could rise above the oceans, it blows carbon dioxide and water back into the atmosphere. By providing a mechanism to return carbon dioxide whence it came, it completes the carbon dioxide cycle. Now we have a feedback mechanism. Amazingly, this will ensure that liquid oceans can remain on the planet, while ultimately reducing the carbon dioxide abundance in the atmosphere today to a fraction of what it is on the other terrestrial planets.

The buildup of carbon sediments on the oceanic crust took hundreds of millions of years, but once a mechanism came into existence to return some of it to the atmosphere, equilibrium became possible. Volcanic activity can feed carbon back into the atmosphere at the same rate it is removed, providing the ultimate feedback loop. The reservoir of carbon in the crust and mantle has become so large over time, for example, that the existing volcanic activity on Earth could replenish all the observed carbon dioxide in the atmosphere in less than 400,000 years. Once the loop is created, and global plate tectonics takes over, the average time for any individual sedimented carbonate atom to flow through the loop before returning carbon dioxide and water to the atmosphere is about 150 million years.

Our oxygen atom fell as part of a carbonic acid molecule in the earliest acid rain in Earth's history, less than 500 million years after our planet had first formed, before significant continental crust formation had begun. It remained locked up in limestone carbonates on the ocean floor for perhaps another 100 million years, getting buried under great masses of limestone as the crust beneath thickened and hardened, making way for later global tectonic movement and the creation of continental crusts. The Earth was now about 600 million years old, and slowly cooling, as the periodic bombardment by meteors and comets had by then greatly subsided. The crust had not yet thickened to the point where global tectonic plates had formed, but at the point where our atom was located, a local form of this process occurred. The mantle was very hot, and convection was churning with a vengeance. A somewhat thickened crust buckled due to the heat and convection currents generated below, and our atom was driven downward as the rock in which it was trapped was subducted toward the mantle, and heated under high pressure.

Here, in microcosm, the first carbon recycling began to occur. At the site of one of the first forming microcontinents, really little more than a volcanic island amid the hot global ocean, about 3.9 billion years ago, our oxygen atom was released in a great flatulent burst. The heat and pressure had expelled gas from the carbonate rock, and eventually the pressure built up sufficiently to drive gas and rock upward to form a volcano, creating new land and shooting out our atom, now part of a carbon dioxide molecule, into the atmosphere.

Through this terrestrial cycle, building up over the next billion years as continental crusts and global plates were established in earnest, our oxygen atom could be exchanged between carbon dioxide, water, and carbonate rocks indefinitely. This eventual 150-million-year cycle of sedimentation, subduction, and volcanic release might completely describe the terrestrial life cycle of our oxygen atom for billions of years to come were it not for new processes yet to occur.

In the first place, there was another, shorter cycle for our oxygen atom to participate in after global tectonics took over. When atoms like our atom resurface at some point as part of the ocean water at what would become the mid-ocean ridges, material from the deep mantle is cycled upward, pushing out and locally melting the crust, creating brand-new material as part of the global plate tectonics game. This material, coming as it does from volatile depleted mantle, does not burst forth with gaseous exuberance, but rather spews in a more controlled fashion out of the mid-ocean vents. At the same time, water is cycled through the vents, percolating down nearby, and being superheated to temperatures in excess of 400 degrees Celsius, and thrust out into the ocean, carrying rich minerals in solution. When this hot water cools upon contact with the oceans, the minerals can sediment out in various chemical combinations. All the water in all the world's oceans circulates through such vents every 10 million years. These hydrothermal vents are therefore rich sources of warm water and mineral solutions in which very interesting chemistry can take place, including the creation of complex organic materials. In early times, when the crust was thinner, the mantle material was undoubtedly driven up in many more locations, so that the early oceans were replete with hydrothermal vents. The recycling of ocean water during that period would have been even faster.

In this way our oxygen atom would lead a full, if predictable, life, ultimately alternating between 10-million-year cycles within the oceans and 150-million-year cycles among the atmosphere, oceans, crust, and mantle. Note that at no time was it yet in the form of molecular oxygen gas, O_2, that we have come to depend on to live. That was still in the future. But remember that each time you breathe in, the oxygen filling your lungs has been a part of the Earth, sea, and sky. It was spewed out of a volcano and rained down upon a steaming Earth billions of years earlier. It may have been locked underground for periods longer than the present ocean floor has now been in existence.

But as our oxygen atom first burst forth from its underground rocky prison, the miracle of chemistry began to work to ensure a far richer future than the preceding picture suggests. Even as the Isua sediments were laid down, the world began to change in a far more profound way than even energetic comets and geological forces could change it. New chemical factories would soon spring up on the planet to completely alter its landscape and atmosphere. They will ultimately recycle our atom 1 million times as fast and 1 million times as often through the environment as would otherwise have been possible. These factories will operate for almost all the rest of our atom's time on this planet.

13.
THE DANGEROUS ENERGY GAME

It is mere rubbish, thinking of the origin of life, one might as well think of the origin of matter.

*CHARLES DARWIN, LETTER TO
JOSEPH HOOKER, 1863*

From the time I was a child, the story of Antoine-Laurent Lavoisier has haunted me. Lavoisier was a scientist in late-eighteenth-century France — considered by many to be the father of modern chemistry — who had the misfortune to be born wealthy and well connected. Of course, had he not been born wealthy he would not have had the luxury to pursue his love of science. Following the family tradition of studying law, he held a variety of public offices during his life. Because he was a member of a consortium that helped the government gather taxes during the French Revolution, he was arrested, and was guillotined in 1794 at age 50 for this capital offense.

The real tragedy of Lavoisier's death, like most untimely deaths, is that we will never know what he might have done. During his life he revolutionized the field of chemistry, creating analytical methods and identifying elements. When he was arrested and tried, he was in the middle of several important experiments, or so he told the judge as he accepted his death sentence, asking that it be delayed until after he had finished this work. The judge, in a statement which should be remembered so that history will never see it repeated, is reported to have claimed "The new republic will have no need of science and scientists. Off with your head."

There are times, such as when the state school board in Kansas in 1999 removed evolution from its science curriculum, when I am reminded of Lavoisier, and shudder at the damage that can be done by ignorance combined with power. Even the magnificent modern edifice called science, built up over half a millennium of small increments toward the truth, is not safe from the vicissitudes of the political world. If, as Carl Sagan claimed, science is a "candle in the dark," banishing the demons that haunted the benighted eras of mankind, it burns tenuously at best. One generation of ignorance, steeped in myth and mysticism, is all that may be needed to snuff it out.

It is the understanding of how a candle burns, in fact, that may have been Lavoisier's most important contribution to the development of modern science. He discovered that the process of combustion did not involve the release of some mythical, spiritual substance, known as *phlogiston*, but rather that the "vital" gas involved was an element he named *oxygen*, from the Greek for "acid-generator," because he thought the products of combustion were always acidic.

Lavoisier did more, however. He showed that humans and guinea pigs alike are merely, in essence, slow-burning candles. He showed that the act of respiration is not designed to cool the body, but rather to warm it, and that the warmth comes from oxygen. In a carefully controlled set of experiments, Lavoisier, along with the French mathematician Pierre Laplace, demonstrated that the heat generated when a guinea pig takes in oxygen and produces carbon dioxide is precisely the same as that generated by burning charcoal. As Lavoisier put it in 1783, "Respiration is . . . a combustion, admittedly very slow, but otherwise exactly similar to that of charcoal." Moreover, following a decade of experiments on human subjects that confirmed the generation of heat from oxygen, Lavoisier ultimately pronounced that "Life is a chemical function."

All the same, as my father used to say, life is hard. There is evidence that all extant life is derived from a single origin during the entire 4.5-billion-year history of our planet. All living creatures use precisely the same molecules to store and transport energy, and ultimately to reproduce. We humans emerged from the same spark of vitality as bacteria and cabbage.

Indeed, once before, as a result of some poetic license, our atom found itself in a world similar to ours. It was just the right distance from a star of just the right size, and was created just long enough after the Big Bang for elements like carbon and iron to have been produced in stars. Yet life there was not to be. Would life have evolved on that long-since-dead planet had not the galaxy intervened? We will never know. After all, even with myriad difficulties, within 100 million years of the time that the laws of physics allowed life to form on Earth, after our planet had cooled and after meteoritic and cometary bombardment had slowed, self-replicating creatures began to change our world. Moreover, these creatures were composed of amongst four most abundant elements in all creation: hydrogen, carbon, nitrogen, and of course oxygen.

If the life later measured by Lavoisier is a chemical function, the prime mover is oxygen. Yet, paradoxically perhaps, oxygen, free oxygen, presented as big a threat to the initial evolution of life on our planet as anything else. Had our atom and all its oxygen cousins existed free and unfettered in the early Earth, life on this planet would probably not have developed at all.

Oxygen, you see, is dangerous, like playing with fire. It combusts. Once combustion has taken place, as anyone who has looked into a hearth after all the embers of a dying fire have gone out, there is not much left to work with. If organic materials combined indiscriminately with oxygen early on, they would have burned up, and would never have been able to build the structures necessary to create the first germs of reproductive life.

Nevertheless, gaseous oxygen is ultimately finely tuned to provide energy for life, beyond its most rudimentary form. While energy can be had elsewhere, nothing burns like oxygen does. Moreover, three minor wonders of chemistry make the molecule of gaseous oxgyen, di-oxygen, O_2, particularly suited to be the source of life's energy: (1) It can release large quantities of energy when binding with other atoms. (2) It takes a lot of energy to get this binding process started. (3) The products of this binding, ultimately carbon dioxide and water, are not themselves highly reactive. Without the first property, complex life could not fuel its existence. Without the second, all organic materials would combust before they could live. And without the third property, life would not survive the process of respiration.

But our oxygen would not know respiration for perhaps 2 billion years after its arrival on Earth. Instead, as we have seen, the atmosphere of the planet was initially dominated by carbon dioxide. Today that gas, produced in abundance by volcanoes, can be lethal. In 1986 it poisoned a whole village of people in Cameroon as it emerged from a lake that had formed above an old volcanic crater.

Somehow, the atmosphere of the Earth proceeded from being dominated by carbon dioxide, followed by nitrogen, to one dominated today by nitrogen and oxygen. We have seen how natural geological processes can and did remove carbon dioxide from the atmosphere. But gaseous oxygen is not the product of inorganic chemistry. Only life could have prepared the Earth for our existence.

After first raining down on the Earth in a water molecule, our oxygen atom became part of a carbonate rock buried for over 100 million years before being released once again to the atmosphere in the form of carbon dioxide. Yet it was still just in time to participate in a wondrous event. Even when the steamy acid rains of the newly forming planet first began to pour over 4 billion years ago, other processes had begun to lay the foundations for the miraculous chemistry we call life.

All life contains the same basic organic ingredients. It was originally thought that perhaps these were created in the atmosphere, by the action of lightning and solar radiation on materials such as methane and ammonia thought to be prevalent in the atmosphere of the early Earth. A famous set of experiments done in the 1950s demonstrated that electrical discharges in closed containers with methane and ammonia did indeed produce complex organic building blocks of life, including the amino acids which are common to all life-forms.

There is a problem with this scenario, however. An atmosphere of methane and ammonia would have been unstable in the presence of the intense ultraviolet radiation coming from the sun. In any case, as was described earlier, any primordial atmosphere was probably short-lived, and quickly replaced by one dominated by carbon dioxide and nitrogen gases.

As we have seen, our atom, along with a significant fraction of the carbon dioxide on Earth, arrived during the latter stages of the Earth's formation, shepherded by comets. Since we now know that complex organic materials are synthesized in comets and meteors, could the same comets that sterilized the Earth during 300 million years of bombardment, perhaps even destroying nascent life in the process, have also delivered the very building blocks for later life to form? In this modern form of the panspermia theory mentioned earlier, life itself may not have traveled through space on comets to ultimately colonize the Earth, but the raw materials that make life possible were delivered in this way.

On first thought, this seems impossible. The impact of a large comet or meteor on the planet is immense, generating enough heat to destroy any complex compounds it carried that survived its interplanetary or interstellar voyage. And as the atmosphere of the Earth built up, heat from friction on such extraterrestrial objects, traveling at velocities greater than 40,000 kilometers per hour, would turn the comets into fiery balls even before they hit the ground.

But not all cometary material falls to Earth inside of the comet. The huge comet tail, visible at night as comets enter the inner solar system, inspiring poets and generations of astronomers, represents the outgassing of material heated by the sun's rays. Small grains of material from the comet can actually be slowed in the upper atmosphere and float to Earth in a gentle fashion. Indeed, such cometary dust has been captured as it falls through the upper stratosphere by NASA U2 planes, and it is replete with organic material. The late Carl Sagan and his colleague Chris Chyba have argued strongly that the organic basis of life was delivered in this manner from space. Moreover, they stressed that if this material was delivered to Earth, it would have also been delivered throughout the solar system, perhaps to jump-start life elsewhere.

One important bit of evidence that supports this contention comes from the Murchison meteorite, discovered in Australia in 1969. This meteorite was replete with amino acids, the building blocks of proteins. As I described earlier, amino acids are complex molecules that can have a "handedness." Like people, they may not be identical to their mirror images. What is particular to life on Earth is that it utilizes only so-called left-handed amino acids in the synthesis of biological proteins. Organic materials that are created in the laboratory, however, usually contain equal numbers of left- and right-handed molecules. While the Murchison meteorite contained both left- and right-handed amino acids, a def-

inite excess of left-handed molecules was found to exist, and the likelihood that this asymmetry resulted from a terrestrial contamination,
while not completely ruled out, is not strong. Could extraterrestrial organic seeds, hitching a ride on comets of the type that delivered our oxygen atom to Earth, have created the eventual built-in preference of
terrestrial life for left-handed molecules?

Of course, even if comets did deliver organic materials in bulk, the
primordial planet Earth with its warm oceans and rich geology had ample other opportunities to create the necessary raw materials for life from
scratch. The idea that life originated in the oceans also has a long history. After all, we live on an ocean-dominated planet, and water nour-
ishes life. A universal solvent, it can transport, in solution, minerals and
volatiles such as carbon and oxygen. All that is needed is an energy input
to cook the stew. The early oceans were hot, and the huge tides washed
water into tidal pools in which the raw materials of life could have been
exposed to radiation that may have driven chemical reactions to make a
host of materials. But the most promising potential locations for the primordial organic factories that may have created our original ancestors
were not even known until about 20 years ago. Only then did the sub-
mersible craft *Alvin*, exploring 2.5 kilometers below the surface of the Pacific Ocean, accidentally bump into what could, depending on your
viewpoint, be termed either a chimney from hell or the Garden of Eden.

The "black smokers" of the type discovered by *Alvin* belch out sulfurous dark fumes rich in minerals from below, and carry water that has
been heated in the crust to temperatures as high as 1,000 degrees Celsius. These are the hydrothermal vents that are found near the mid-
ocean ridges, where material from the deep mantle flows up to the
surface, creating new crust that drives the continents apart in places and
forces them together in others. Our oxygen atom, trapped in the carbonate rock below the ocean, was ultimately first freed due to the movement of primordial crusts generated by such flows. At that time, as the
mantle material thrusted up through the thin crusts, hydrothermal vents
must have existed in great abundance.

Shortly after these black smokers were found, whole new species of
animals were discovered to thrive around them, from scores of bacteria
to huge tubelike animals that feed on them. The bacteria feast on sulfur

and some are poisoned by oxygen. In this dark midnight world, we may find the remnants of our earliest ancestors.

In a cosmic sense, the story of life, which will enrich the lives of our atom a millionfold, follows closely all of our atom's previous history. If the universe had remained in equilibrium, everywhere, for its entire existence, nothing of interest would have happened. Instead, from the creation of protons to the creation of stars, we have witnessed over and over again a local departure from equilibrium, followed by a return back to the fold. Everything of interest to us in the universe has followed from these momentary deviations. Protons exist only because of a departure from equilibrium that allowed matter to stave off annihilation with anti-matter. Following that, gravity became the engine for much of the action. Small density fluctuations grew, and led to localized storehouses of energy. The gravitational energy of falling matter was converted into heat energy, which powered the stars, keeping them locally in equilibrium, but fighting a losing battle against the inevitable collapse which will pour that energy back into the universe. As a result of this storage of energy, the elements that make life possible were created. Departures from equilibrium in molecular clouds shielded elements like carbon, oxygen, and hydrogen on the surfaces of grains from radiation that would have stopped them from combining into complex molecules. And ultimately life itself exists only as long as it can perpetuate departures from the inevitable sharing of energy and disorder that govern the universe as a whole. Darwin was right: From an underlying physical perspective, the origin of life and the origin of matter are very similar. The interesting question, which we shall return to, is whether their end is also the same. For the moment, however, we are concerned with the beginning.

Life can exist only if it can acquire energy from its surroundings, more energy than is its due. Life represents order in a universe that is designed for disorder. And as much as we may debate where the ingredients for life first arose, these ingredients alone do not make life any more than the components of an engine make a car a car. The spark of life and the fuel to keep it going animate the inanimate.

If energy is the engine that drives life, the source of this energy is the electron. Life consists, at its most basic level, of a self-replicating mechanism for transferring electrons, by dividing and combining molecules, to produce useable energy. Electrons from energy-rich molecules are stripped away and pass through biological systems until they no longer have anything to give. They are passed on to the environment, and new electrons are found. All living systems obey this simple rule. They may differ in where they obtain their energy, in how they use it, and how they dispose of the waste electrons, but these differences are minor in the overall picture. And because the gift of manipulating energy-rich electrons runs so counter to the normal process of physical systems, which give up energy readily rather than hoard it, we only have evidence on Earth that this trick was learned once. Every living thing on Earth over the past 3.5 billion years, as far as we know, has adopted precisely the same mechanism for storing and transporting energy obtained in different ways from the environment. The mechanism, once discovered, is remarkably robust. As far as we can tell, life exists everywhere on Earth where there is a chemical reaction that can provide a source of energetic electrons as fuel.

The mechanism seems to have been discovered within 100 million years or so of the time the Isua sediments were being laid down and continents were beginning to form — about the time our atom was released from its long sojourn underground, as plate tectonics began to take over the dynamics of the crust. At this time, microcontinents of the sort that our atom emerged from poked their heads above the water sporadically. These volcanic islands were no doubt surrounded by hot springs and tide pools. Perhaps it is here that life was first created, or perhaps it was deep underwater, closer to the bowels of the Earth.

We will return to this mystery shortly. Let's instead now focus on the evidence that life indeed first took hold early on, as our oxygen atom was released from its incarceration. This is a very important recent discovery, because it may have profound implications for the possible origin of life elsewhere in the universe. Life seems like a remarkable accident, but if it was an accident that could occur in less than 100 million years, given the proper conditions, this vastly broadens the possible creation sites throughout the universe, and perhaps even throughout the solar system.

For a very long time, it was thought that the history of life on Earth was restricted to the past 500 million years or so, starting in what has become known as the *Cambrian era*. The reason was simple: Fossil shells and skeletons were found dating back to that time, but not before. However, as Carl Sagan was fond of saying, absence of evidence is not necessarily evidence of absence!

And indeed, absence there wasn't. As often happens in science, because the prevailing view was that there was no life before the Cambrian not a lot of effort was expended to find it. But while Precambrian fossils were first described in 1899, they were finally generally recognized to actually *be* Precambrian about 50 years ago, following various discoveries in quick succession around the globe. Microfossils were discovered in the Gunflint Chert around Lake Superior in 1952, some of the earliest Precambrian animal fossils were found by an English schoolboy walking in the a forest in Leicestershire in 1957, and others were identified in the Flinders Mountains in southern Australia in the 1960s. After that, a plethora of examples soon turned up from Russia; so did the famous Burgess Shale in British Columbia. It was not surprising, in retrospect, that these prehistoric animals had previously escaped detection. They were soft-bodied, and thus did not bequeath shells or skeletons to the future.

But the march of life did not start there. While multicelled animals appeared just before the Cambrian, multicellular algae-like material has been dated back at least 1.5 billion years. A primitive form of apparent multicellular organic matter uncovered in China is dated to be almost 2 billion years old.

Two billion years may seem like a remarkable unbroken stretch for life to exist, but it barely scratches the surface. The further back our techniques allow us to explore, the further back we find life. Indeed, we now know, both from direct and indirect evidence, that the murky origins of life might be found in the oldest rocks themselves, almost 4 billion years old.

Indeed, the rocks at Isua may have contained the seeds of our existence. These sediments harbor carbon polymers, which certainly could have been created by prebiological chemistry. Two isotopes of naturally occurring carbon exist in nature: carbon-12, the dominant isotope, and the slightly heavier version, carbon-13. Living systems have a slight preference for utilizing carbon-12. The carbon in the rocks at Isua has had a complex history, having been subjected to pressures as high as 5,000 times atmospheric pressure, and temperatures of about 550 degrees Celsius. These conditions produce some transformations that mix up carbon in the rock with material from other sources. Nevertheless, the Isua rocks do show some depletion of carbon-13, which some researchers have interpreted as evidence that living organisms may have created the polymers. It turns out, however, that there is also a significant predominance of carbon-12 over carbon-13 in the carbonaceous chondrite meteorites which were bombarding the Earth with high frequency up to that time. Thus one cannot rule out the possibility that the observed asymmetry might be a remnant of extraterrestrial bombardment.

In all sedimentary material less than 3.5 billion years old — namely, material that has not been significantly processed inside the Earth — the signature is far clearer that organic life had definitely begun to thrive. Here, there is no mixing of materials, and the observed depletion of carbon-13 is precisely that which is observed in known organic, that is, living, systems.

There is a tenuous piece of evidence that life may have emerged earlier than this, back when microcontinents first came into existence, and it again comes from the Isua rocks. In these early rocks, phosphates can be found. Phosphorus is an extremely reactive and toxic substance, but it is tamed when bound to 4 oxygen atoms, and in this form it is manufactured in all living cells today. Some have argued that the presence of phosphate compounds in the Isua rocks suggests that life was already in existence at this time. It should be noted, however, that while phosphates are manufactured by living cells, they are not manufactured *only* there. Phosphates, for example, occur on the moon, and no one suggests that this is indirect evidence for life there.

What makes phosphates worth focusing on, however, is that life, as we know it, and phosphates are intimately tied together. Living systems have many different ways of borrowing energy from the environment, but they have developed only a single way of manipulating this energy to create organic matter. Every living system ever discovered, from the lowliest bacteria to Albert Einstein, whether it gets energy from the sun, from eating other organic material, or from the sulfurous gases that belch out of the Earth, relies on a substance called *adenosine triphosphate* (ATP) to survive. This material has three different phosphate groups on a chain connected to a carbon ring. The bonds connecting the outer two phosphate groups in the chain are *high-energy bonds*. Breaking off a phosphate group can therefore release tremendous energy. Breaking off the outer phosphate results in *adenosine diphosphate* (ADP), and breaking off the second produces, as you would no doubt guess, *adenosine monophosphate* (AMP). The energy gain from such transformations can be used to build an animal capable of living and reproducing, and, I repeat, it is the *only* way that living systems on Earth have, for the past 4 billion years, redistributed energy to where it is needed. The presence of phosphates amid the remnants at Isua is by no means definitive evidence that life existed there and then, but it gets one wondering.

This brings us to the inevitable chicken-and-egg question. It is reasonable to expect that ATP existed in the environment before life had developed to synthesize it. Experiments have shown that ATP can be naturally created as complex organic compounds are built from compounds on comets and meteorites, and that also would have existed in the prebiotic Earth's atmosphere, oceans, and rock. It is therefore likely that the earliest forms of life simply utilized ATP by harvesting it readily from its environment. This could not go on forever, of course, and eventually an evolutionary advantage was obtained by those life-forms that could synthesize ATP within their bodies, using other energy sources from outside as fuel.

So the phosphates within the Isua sediments may have been biologically created, or may have been the fodder for life that had yet to form. Either way, we are now close to that magic moment. And we can get a clue to where that occurred by searching for definitive evidence of primordial life, the fossil remnants of the earliest life-forms themselves.

In science fiction movies, whenever explorers uncover some odd, often threatening, primordial ooze dating back to the dawn of time, it is most often in a wild, remote location. Why does this seem more natural than, say, downtown Milwaukee? A moment's thought gives the answer. Life is forever in the act of reinvention. And while it is perhaps not profound to point it out, hospitable locations are hospitable. They imply easy living. Easy living draws newcomers, and newcomers inevitably displace the original inhabitants.

So if we want to turn back the clock and seek out lands that time forgot, this reasoning suggests we go where no one but determined explorers might choose to venture. Only then can we expect to find hardy survivors of antiquity. Also, perhaps because life, especially complex life, invariably alters its environment, similar reasoning suggests other remnants of the past might also find sanctuary in remote locations. Or perhaps it is simply because much of the Earth is still poorly inhabited that the oldest rocks and fossil remnants known to humans have been discovered far from the beaten track.

If Hollywood were to script a movie in which prehistoric life is discovered still thriving in a remote location, the continent of Australia, separated from the other continents 200 million years ago, would not be a bad choice. Australia's outback contains some of the most inhospitable locations one can imagine. I have been told by researchers who have worked there that while not everything that moves is poisonous, it is not a bad assumption to make. The vegetation is not much more friendly. The paleontologist Richard Fortey has written eloquently of surroundings where "every shrub is equipped with spines, and those that are not, are equipped with burrs."

And so it is that Australia harbors not only fossils from among the oldest living creatures known on Earth, but their largely unchanged descendants today. The fossil remnants are not immediately distinguished by the individual shapes of the beasts who died to produce them, but rather by their resemblance to large, green, tacky pillows made of phyllo dough.

In western Australia, about 400 miles north of Perth, in the salty tide pools of Shark Bay, are found such collections of *living* objects, called *stromatolites*. These are literally colonies of microbial life, with different metabolic paths, living in harmony. Slicing through the layers of these objects is like reading through a history book of life. The creatures on top, called *cyanobacteria* (blue-green algae), live off the energy of the sun and carbon dioxide, and obviously can survive in the presence of the oxygen in our atmosphere, and even make some use of it. These little guys, less than one ten-thousandth of an inch in size, contain the molecule *chlorophyll*, which allows them to absorb light and use the energy to break down carbon dioxide into carbon, for their own organic nourishment, and oxygen, which is released as a waste gas to the atmosphere.

Below this layer, other bacteria thrive. Immediately beneath the top layer are bacteria that can exist in concert with oxygen but function perfectly well without it, producing energy by fermentation of the waste products from the bacteria above them. Below these objects are bacteria that are completely *anaerobic*, that cannot tolerate oxgyen's presence at all. These bacteria thrive on the waste products of the layer above them, and are also nourished by the minerals that lie in the grains of sediment trapped by the uppermost layer of bacteria, and which remain with their dead bodies as the stromatolite structure grows upward. In this way the lower bacteria constantly feed on the dead bodies of their upper-story neighbors.

Over 100 years ago, the paleontologist James Hall discovered what are now recognized to be fossil stromatolites, layered patterns of rock formations, containing what appeared to be microscopic fossils. Great debate raged for almost a century as to whether these were indeed organic, but eventually in the layers indisputable microscopic fossils, almost identical in form to modern-day cyanobacteria, were finally discovered. Under the microscope they resemble miniature turds, less than 50 millionths of a meter in length. They have been preserved in the silica paste that filled the inside structure of the prehistoric ancestors of the modern stromatolites.

Here are objects that have lived, almost unchanged, for much of the Earth's history. The oldest known fossils are found — where else? — in western Australia in the so-called Pilbara rocks near the pleasant-sounding location of Warrawoona. But equally old remnants are found in the Fig Tree rocks from South Africa. These have been dated to be 3.5 billion years old.

The debate that originally raged as to the organic nature of the stromatolite fossils was not misconceived. Some people may argue that scientists are too conservative, automatically rejecting evidence of new and exciting possibilities. But it is important to realize that most new and exciting possibilities turn out to be wrong. A modern version of the stromatolite debate erupted recently over the possibility that a 4.5-billion-year-old rock, not from Earth but from Mars, contains fossils smaller than but similar to those first seen in the stromatolites. Tiny turdlike formations have been observed in this rock, and some materials sometimes associated with life have been measured within it.

As exciting as this is, because it would represent the first observation of extraterrestrial life in any form, even if it is long dead, some caution is advised. The evidence of fossils in this meteorite found in Antarctica, with the romantic name ALH84001, is much more tentative than it was for Hall's stromatolites. In particular, recent analyses suggest that the organic-looking formations were created at temperatures that were far too high to have accommodated life. Considering that it took many independent samples obtained over a century to settle the debate about stromatolite fossil life in prehistoric Earth, you can bet it would take many samples and more definitive evidence before most scientists would be convinced that life, even microbial life, existed on Mars. This is as it should be.

Returning to Earth, another debate is now looming over whether recent discoveries, again in Australia — this time three miles below the seabed in a petroleum exploration well — represent not fossil life, but living life! Indeed, as I was writing this, a report appeared describing tiny *nanobes*, strands of organic material billionths of a meter in length, that grow between the mineral crystals in the highly compressed sediments at this depth. They respond in some tests as if they contain DNA (although DNA has yet to be explicitly extracted from them), and reproduce quickly in dense colonies of tendril-like structures clinging to rocks. These are easily as small as the Martian fossils, if not smaller. In fact, they are, according to some biologists, too small to represent fully blooming life. Will these stretch what we mean by "life," and could they represent a link to the earliest missing links between life and organic molecules? Stay tuned.

In any case, stromatolites appear to have flourished from at least 2.5 billion years ago until about 500 million years ago. Thereafter, they declined; fossil remnants from this period include individual single-celled microbes found throughout the world, from the Arctic islands of Spitsbergen in Norway to the rocks of the Canadian Shield. The reason for their later decline is clear. As long as there were no living predators, the bacteria could thrive. As soon as more complex life-forms arose that could eat the bacteria, their heyday was largely over. That is, except in places like Shark Bay, or in the Baja in Mexico, where heat combined with incredibly high levels of salt make life unbearable for anything but the hardy descendants of these early creatures (and, of course, the hardy scientists who explore there).

The cyanobacteria in stromatolites have free-floating cousins, known as plankton. They range from a millionth of a meter in size to 20 times this value, with the smaller critters living deeper below the surface, down to about 80 meters. Plankton today account for almost 80 percent of the oxygen produced in non-Arctic waters.

If the earliest fossils found are indeed remnants of bacteria that utilized light to produce oxygen, then life had already advanced substantially by 3.5 billion years ago. For almost certainly this was not the original form of life. The complexity required seems too great. Moreover, as we shall see, the direct production of oxygen by photosynthesis would have been too traumatic. That would have been like humans breathing out lethal cyanide gas in the process of respiration. In addition, without oxygen in the atmosphere and its consequent ozone layer, ultraviolet radiation would have been a powerful killer. Best to avoid light altogether.

Instead, the first cells were probably more accustomed to the darkness and the putrid smell of sulfur associated with hydrothermal vents. Every year one reads of new forms of life discovered in places ranging from the relatively benign hydrothermal vents to the acidic, toxic, sweaty regions at the bottom of deep oil wells.

Most compelling of all, perhaps, is the recent discovery that the tree of life has merely three branches, not five, and that the one closest to the root involves bacteria that live in hot environments, the *hyperthermophiles*.

Hyperthermophiles defy all conventional wisdom. These forms of life not only can thrive in environments that normally sterilize materials, in excess of the normal boiling temperature of water at sea level, 100 degrees Celsius, they *require* such temperatures. They die if it is not hot, and many cannot reproduce if the temperature drops below 80 degrees Celsius. They can eat sulfur, and in the presence of oxygen many die.

The understanding of life has changed dramatically as we have unraveled the details of the genetic code governing its replication. At the center of this process is DNA, the long double-helix molecule with its digital information contained in four nucleic acid pairs that can connect the two helices in two ways, either by a cytosine-guanine (CG) bond or a thymine-adenine (TA) bond. The sequence of ATCG groups on the molecule uniquely determines the genetic code that in turn determines every single aspect of the resulting life-form.

The field of molecular genetics has completely changed the way we compare animal species. I vaguely remember that when I was a student, I was forced to memorize the five kingdoms of life: plants, animals, fungi, bacteria, and protists (sophisticated single-celled animals). One of the gratifying aspects of the march of science is that, at least in principle, there is less and less to memorize the more we understand. This is because more of the ultimate complexity of the universe can be understood as following inevitably from a few fundamental principles. (Of course, I will get in trouble here with a few of my colleagues who resent this "reductionist" view of the world, and argue that new laws emerge to describe complexity on ever larger scales. But this is an auxiliary debate which should not detract from the key point that the more we understand, generally the less we have to memorize.) Five kingdoms have now become three branches. And the three branches involve simply two types of basic building blocks. Two of the three branches of life include various types of bacteria, with single cells that are basically just sacs of DNA and organic materials shielded from the outside world by a cell wall. These are called *prokaryotes* (pre-nucleated cells). The other, more advanced, if perhaps more fragile, form of life, *eukaryotes*, have cells with nuclei containing the genetic material, and all sorts of other stuff whose names invariably cause my mind to glaze over. There is one item, though, that is particularly relevant, and provides yet another piece of evidence, if one were needed, that these cells arose later than prokaryotic cells. These cells contain structures called *mitochondria*, where oxygen is used to burn food for energy. In an early world without free oxygen, there would be no need for these objects.

What is particularly interesting about the prokaryotes is that they are just simple sacs of chemicals that reproduce. Crucial to their survival are the cell membranes that control what comes in from the outside world.

Surprisingly, such membranes can arise spontaneously amid certain organic materials that have been found on carbonaceous chondrite meteorites. Indeed, when extracted from the meteorites they have been observed to congeal naturally into membranes surrounding sacs called *lipid vesicles*. This housing for the chemical factories that we call life could thus have naturally populated the ancient oceans of Earth, resolving one big step on the road to existence.

Simpler than DNA is the single-stranded RNA, which is basically half a DNA molecule with the T's changed to U's (uracil) and an extra oxygen running along its backbone. RNA has a variety of different uses, including transmitting genetic information copied from DNA from one place to another and helping construct proteins. Because it is simpler, and because of a recent Nobel Prize–winning discovery that some forms of RNA can actually catalyze reactions that can splice and rebuild RNA molecules, it is now believed that prior to the complexity of our DNA world, an RNA world existed. Here this molecule was used to build and convey genetic information.

Most important, for our purposes, are the *ribosomal* RNA (rRNA) molecules, the ones that encode information to create enzymes necessary to catalyze chemical reactions used by all living things. One particular rRNA molecule has been studied in some detail, as it is common to all living things. This is the 16/18s rRNA molecule, which sounds like a tire size. (The 18s sequence exists in eukaryotes, and the 16s sequence in prokaryotes.) This sequence is relatively simple by DNA standards, and contains about 1,500 bases.

What is perhaps most striking is that all species in existence have the same types of these molecules, suggesting that we all have a common ancestor. Of course, the detailed sequencing of the bases is different in different species. But as different as they are, they are all recognizably related. Thus while people don't bear a great resemblance to bacteria (well, some people do, but I would rather not name names), they are not so different that their 18s rRNA cannot be identified with the 16s rRNA in bacteria by finding large common areas among the sequences that are almost identical.

Every now and then the genetic copying process produces an error in the 16s or 18s sequence. If this error occurs in a part of the sequence that the host cell relies on, it will either produce a nonviable entity (usually), or a better model. In this case, natural selection will influence how fast this error propagates. There are lots of parts of the sequence that have no apparent function, however, and appear to be essentially redundant. If one examines "errors" (random changes) in these nonfunctional parts of the sequence, one finds that these occur at a fixed rate, without being influenced by natural selection because the changes produce no effect to be selected. Instead, the rate of such changes is simply governed by the inherent rate (which may vary over time) at which random errors result in the genetic copying mechanism. The more similar these sequences are between species, the more recently they diverged on the tree of life.

And it is thus that a *phylogenetic* tree of life has been developed by biologists (who like to add Greek- and Latin-based adjectives to every noun). It has three branches: the *eukarya* (containing cells with nuclei, thus including all present-day plants and animals), the *bacteria*, and another branch called the *archaea*, which are also prokaryotes, but differ substantially from the bacteria in the structure of cell walls and other properties. Almost all the archaea are hyperthermophiles, as are some of the bacterial species.

As the name *archaea* suggests, it is believed these species are truly archaic, in the sense that they predate the eukaryotes, and moreover that the eukaryote branch (and perhaps the bacteria branch) diverged directly from this branch on its way to creating us. This inference is not ironclad, however. After all, while the technique of examining ribosomal RNA can demonstrate how close, evolutionarily, two species may be, it cannot directly determine which came first. But the fact that all species contain the same types of rRNA molecules suggests they shared a common ancestor. Moreover, a look at the divergence between rRNA sequences suggests that eukarya and bacteria are related more closely to archaea than to one another, suggesting, perhaps, that the latter may have contained the common ancestor. If you trace your own family tree back far enough, you are thus likely to find you are related to a sulfur-eating bacterium!

These arguments suggest that all life on Earth today descended from species that liked it hot. Remember that the archaea are by and large hyperthermophilic — they thrive in hot water, such as that near hydrothermal vents in the ocean floor. Furthermore, many of them are anaerobic, and may only survive in environments in which free oxygen is not present.

This doesn't imply that life itself began in such environments, only that it went through such a stage, and everything around today stems from those life-forms. It does stand to reason, however, that present-day life evolved out of bacteria that thrive without oxygen, perhaps without light, and only in hot water. In the first place, in the early Earth there was no free oxygen. Next, in the absence of oxygen, there was no ozone layer to protect life against the extreme ultraviolet radiation coming from the sun. While there is little doubt that life can survive such conditions, this may nevertheless have inhibited its growth on the surface of the oceans and on land.

We must also remember that the cometary and meteoric impacts so prevalent before about 4 billion years did not stop all at once, but rather tailed off gradually. It is almost certain that more than one ocean-evaporating collision, requiring an impact on the Earth by an object more than 300 kilometers in radius, occurred prior to about 3.8 billion years ago. But following this time, perhaps up to about 3.5 billion years ago, impacts by smaller objects, perhaps 150 kilometers in radius, are not statistically out of the question. These would effectively have evaporated all of the world's oceans down to a depth of 200 meters. Thus if surface-living species had evolved up to that point, they could easily have been destroyed by such events. Hyperthermophiles, however, living near the ocean bottom, could have survived unscathed.

Finally, biochemical arguments suggest that sulfur-eating bacteria, or methane-producing fermenters, are likely to have predated more sophisticated photosynthetic bacteria. In fact, before photosynthesis there was quite likely chemosynthesis. Here, primordial life-forms would have lived without oxygen and in the dark. They would not have been powered, as plants are, by the sun, but rather by the heat of the Earth.

Remember that the motor that drives life is simply based on the movement of electrons. Certain elements are electron donors and others are receivers. The process of receiving the electrons can liberate energy, which is then used for other purposes. Hydrogen likes to donate electrons, and oxygen likes to take them. Whenever an electron is added to an atom, its overall positive charge is reduced, since electrons are negatively charged. Such an atom is said to be *reduced* by this process. Since oxygen likes to accept electrons, the materials it removes electrons from are called *oxidized*. This term is more generally applied any time an electron is removed from a system. As a rule, whenever materials are oxidized, energy is released. Burning is a good example of this: Here, carbon compounds are oxidized by oxygen gas, producing fully oxidized carbon in the form of carbon dioxide, CO_2, plus H_2O. Alternatively, when materials are reduced, this reduction usually takes in energy, and this energy is then stored for further use.

In the early, hellish Earth, noxious fumes and various mineral combinations would have been spewing in abundance at the hydrothermal vents and hot springs dotting the ocean floors. The high temperatures in the Earth would have contributed to the formation of various reduced compounds which could then have provided the fuels for the earliest life, by oxidation. The process in which an iron–sulfur compound FeS oxidizes the reduced (and very toxic) gas hydrogen sulfide (H_2S) to liberate energy (and electrons), producing in the process hydrogen ions and the metal iron pyrite (FeS_2), has been suggested as the first source of energy used to build the organic compounds called life.

What makes this process particularly interesting is that it is completely inorganic, and therefore does not require the pre-existence of complex organic compounds. But more important, the structure of iron pyrite is quite regular, with positive-charged iron sites to which organic compounds can bind. It has been suggested that iron pyrite might thus have served as a template on which large, regular organic polymers such as RNA may initially have been formed. In this way, the hydrothermal vents would have provided, by their heat energy, both the source of energy to fuel life and the structures that could have helped to synthesize the complex compounds like RNA that allow life to replicate. In fact, hyperthermophilic iron-reducing bacteria continue to exist today.

A plethora of other kinds of hyperthermophilic archaea bacteria exist that glean energy from similar inorganic donors of electrons, making use of the gases produced near the vents, including pure hydrogen gas and hydrogen sulfate. The former bacteria produce methane gas, while the latter produce hydrogen sulfide, both pretty disgusting and generally toxic to life like us. All of these, based on the rRNA analysis described earlier, fall near the root of the tree of life, though they have persisted for almost 4 billion years. And some of them are quite similar to bacteria found at the base of modern stromatolites.

Notice two things, however. First, nowhere does free oxygen play a role in any of these processes. It is neither consumed or produced. In a related vein, none of these processes release the standard products of burning, namely, carbon dioxide and water. Instead, they produce energy, which is used universally for a single purpose: to break apart carbon dioxide, so that the carbon atoms can be assimilated into larger organic structures as the organism grows, develops, and reproduces. It is here that our oxygen atom first comes in contact with life.

Almost 4 billion years ago, our oxygen atom emerged from its underground prison as part of carbon dioxide, spewed from an underwater volcano that was just about to rise above the surface of the primordial seas. The carbon dioxide dissolved in the surrounding water, which flowed over hot springs and vents covering the ocean floor. Here, the first microbes were waiting to pounce on their prey. Unlike animals that eat their prey to get hold of the organic storehouse of materials they can provide, our prehistoric microbes ate and breathed things like sulfur for energy, and they ate carbon dioxide for its raw materials. Absorbed through the cell wall of one of these microscopic critters, our oxygen atom, still bonded to its carbon partner, was rudely disturbed while a hydrogen atom was inserted into the mix. Then another hydrogen replaced its neighboring oxygen partner, and our atom now found itself part of what some people lately are trying to avoid eating, a carbohydrate.

Our atom is now a part of an incredibly rich chemistry. Instead of microscopically slow transformations from carbon dioxide to carbonates, to water, to carbon dioxide, taking place over hundreds of millions of years, our atom becomes a part of materials that change their composition in years, days, or hours.

Even these microscopic bacteria predators are already complex machines. They have, at the very least, built up RNA for reproduction and catalysis and ATP for energy storage. Moreover, they must transport the energy, starting out as high-energy electrons, down a chain that eventually leaves lower-energy electrons excreted as waste. The energy released is used in turn to create ATP and other organic compounds, all beginning in this case with carbon dioxide and water.

As organic materials build up, and in fact perhaps even before these chemolithoautotrophic hyperthermophilic processes (I always wanted to be able to say that!) were established, another process can power life. The organic remains built up as the life's work of one microbe can be used to power another. When its host dies and falls to the ocean floor, for example, our oxygen atom, locked in its carbohydrate, can be subsumed as food for another bacteria, which can be powered, again in the absence of light and oxygen, by fermentation.

Fermentation, familiar in wine making and bread baking via microbial living yeast, is simply the breaking down of complex organic materials such as sugar (glucose) into ethanol and carbon dioxide, releasing energy in the process, which can in turn be used to build up other organic materials, in particular ATP. In this way, our atom can become bound up in the phosphate group of an ATP molecule, becoming a tool in the energy machine that powers life.

Fermentation, however, is a relatively inefficient way of producing energy. Of the 6 carbon atoms in glucose, for example, only 2 end up releasing their energetic electrons and becoming fully oxidized as carbon dioxide, and the rest continue to store energy in ethanol. Nevertheless, with sufficient food, and lack of any other raw materials, fermentation will do in a pinch. For example, the bacteria in the middle layers of living stromatolites carry out this process.

Life soon improved on the need for independent processes of fermentation and inorganic oxidation by combining the two processes. One brand of sulfur-eating bacteria, for example, will ferment, but if it is in the presence of sulfates (sulfur–oxygen compounds), instead of simply leaving some of the higher-energy electrons unused it will transfer them to a site where they can release energy. This helps create more ATP, and the leftover electrons are transferred, reducing the sulfates, excreting out hydrogen sulfide gas. In a sense, as we shall see, these bacteria can be said to breathe sulfates, as we breathe oxygen.

In order to gain more useful energy from its food, these bacteria must develop a mechanism to transport the electrons, accepted in one location, to be donated later in another location. The fact that certain anaerobic fermenting bacteria do not have or utilize such a chain suggests that they are among the oldest living species. Eventually, however, microbes hit upon a ringlike molecule called *porphyrin*. At the center of a ring of carbon atoms a single iron atom can be located, in which case this is called a *heme* group. The particular structure of this group allows electrons to flow easily within it. In this way, they can be accepted from outside, move to the middle during transport, and then be redeposited elsewhere.

The development of these structures and of the associated energy transfer and production processes they mediate is of crucial importance for the future of our oxygen atom on Earth. For it makes way for the two most profound developments in the history of life: photosynthesis and, later, respiration. By these two processes, not only would life be forever changed, but so would the Earth.

14

THE WONDER YEARS

Yea, slimy things did crawl with legs
Upon the slimy sea.

SAMUEL TAYLOR COLERIDGE

If we could build a time machine and go back 3.5 billion years, when enough continental crust existed for us to find a rocky shoreline to stand on, the Earth might not look that different from how it looks today. The ocean waves carrying lipid foam would crash against the rocks, which themselves might be covered with slimy scum. But looks can be deceiving. Set down in this archaean landscape, with temperatures possibly hotter than Death Valley in July, you would be dead in two to three minutes. The danger is invisible to the eye. Oxygen is nowhere to be found. Indeed, precisely because of this, gases such as hydrogen sulfide, which otherwise would be quickly oxidized, can survive. It is then a toss-up whether brain death would occur because of carbon dioxide poisoning or the more lethal effects of hydrogen sulfide.

Nevertheless, life had already begun to take firm root on this planet, and the planet has been inhabited continuously ever since. But life was at a crossroads. Two sources of energy existed to be taken advantage of: (1) the heat of the Earth, which produced a stew of energetic reduced compounds boiling up from hydrothermal vents and volcanoes, and (2) pre-existing organic materials that could be cannibalized. Both sources of energy were tenuous, however. As the Earth cooled, the convection flows would slow, and crust would build, so that the abundance of upward-thrusting hydrothermal energy sources would decline over time. And while the early broth might have been rich in organic chemicals either transported from space or created in the primeval atmosphere, this resource too would have had its limits. Moreover, most of the planet would have remained inaccessible to life in such a world. Only in isolated pockets would life thrive. To colonize the entire planet, a new energy source would have to be found.

Our oxygen atom has already experienced both life and death. It was incorporated into complex carbohydrates as bacteria built themselves up, and then when they died it was cannibalized by other bacteria that ate this material and through fermentation produced energy, breaking down the compounds containing our atom back into carbon dioxide. In this way, our atom has been able to maintain a busy existence within the warm ocean water. Nevertheless, it has been, up to this time, merely a pawn in the game of life. All that will soon change.

The first change comes quite naturally. In the atmosphere, the abundant ultraviolet light from the sun is as yet unshielded. This light can, by a process called *photolysis*, break apart individual water molecules in the upper atmosphere into their component hydrogen and oxygen atoms. This reaction requires substantial energy, but there is substantial energy coming from the sun. Soon life-forms develop that can take advantage of this solar energy. And take advantage they do. The groundwork has been laid as living systems have already developed mechanisms to transport electrons on a chain from where they are produced with high energy to where this energy can be released.

The changes taking place on the surface of the planet are now quite visible. Multicolored bacteria, purple, green, and yellow, begin to colonize all available perches on the growing shorelines of the world. The colors are not for show, nor to attract mates or lure animals. Sex has not yet been invented, nor have animals. The process of photosynthesis — capturing light energy from the sun for use in the production of organic material — involves in all cases some form of chlorophyll, a molecule that creates the green color we see in plants. Chlorophyll in turn contains porphyrin rings, similar in type to those we encountered earlier, except that they have a magnesium atom at their center instead of an iron atom. This is one reason it is logical to suspect that photosynthesis was not part of the first metabolism on earth. The sulfate-reducing bacteria,

which in turn evolved from simpler fermenting bacteria, had to develop
the mechanisms for creating porphyrin and other basic components that
would later be used to harness solar energy.

The process of light absorption by any atom or molecule is the same.
Electrons orbiting the atoms absorb the light, and the electron energy is
raised in the process of creating an "excited" atom. Normally, after some
time the electrons release this energy, once again in the form of light.
The colors of light they emit need not be the same as the colors they absorb, depending on precisely how the atom relaxes to its unexcited state.
Moreover, some atoms remain in their excited state for some time, so
that they emit light long after they absorb it. Such phosphorescent molecules are used, for example, in wristwatches that glow in the dark.

A single chlorophyll molecule will normally absorb light and re-emit
it again in due course. But a chain of chlorophyll molecules in a living
photosynthesizing organism can instead harness the energy of the excited electrons and pass them on to the energy transport chains, wherein
they are used to make ATP, which stores the energy for later use by the
organism. In addition, another molecule, called NADP, is reduced (by
the addition of hydrogen) to form NADPH, which also stores energy for
the organism.

All of these reactions proceed following the absorption of light by
chlorophyll, and are thus called *light reactions*. Following these, a second stage of photosynthesis takes place, called *dark reactions*. These reactions make use of the previously stored energy first captured by the
chlorophyll or other pigments. Here, carbon dioxide, such as the carbon
dioxide molecule containing our oxygen atom, is reduced by the addition of hydrogen atoms to make organic materials such as glucose, leaving water as an additional by-product.

Note that throughout these processes, a source of hydrogen atoms is
needed. The earliest photosynthesizing bacteria got their hydrogen from
the same source their predecessors got their energy, the products of hydrothermal vents. Green and purple sulfur bacteria, still found today, get
their hydrogen from hydrogen sulfide, produced volcanically (or by
other biological systems), while other purple bacteria get hydrogen from
organic remnants of their dead ancestors.

Once photosynthesis became possible, life was free to rise to the surface and proliferate throughout the planet. The absence of oxygen meant that reduced substances, that is, those containing available hydrogen, were long-lived, and in addition, the organic material produced by photosynthesis could be used to feed other, fermenting bacteria. Symbiotic colonies, early ancestors of the colonies that create modern stromatolites, began to populate shorelines. In these colonies, photosynthetic bacteria would form large mats — like primitive solar cells — that captured radiation and converted it to organic material. As they died, their bodies would build up in layers that could then feed fermenting bacteria located below them.

These processes of life and death then began to take place with full force. The buildup of organic materials would serve to remove carbon dioxide from the atmosphere. If it were later fermented, some of this carbon dioxide could be returned to the atmosphere. Some of the dead organisms, however, would simply merge with the rocky sediments that would build up around them. This organic carbon would then become trapped in the continental or oceanic crusts, to be recycled only over geological timescales. In certain places where crustal remnants from this period are found today, carbon-rich deposits exist in seams that are comparable to the later coal seams created by tropical forests billions of years in the future.

Thus even at its earliest stages, life was already beginning to change the Earth's environment, both above ground and below. Over billions of years, the natural processes of life and death have contributed to the overall removal of carbon from the Earth's atmosphere. It is estimated that 20 percent of the available reservoir of carbon on Earth, including that bound up in the crust, has passed through living systems over the past 3.5 billion years.

The same processes that induced life inspired some of the early deaths. The energetic ultraviolet radiation from the sun raining down on Earth can just as easily break apart molecules as provide them with energy. Undoubtedly some of the primordial bacteria exposed to this light were killed by it. It is quite likely, for example, that in some of the primordial stromatolites, the upper layers died relatively quickly. Below them other layers of photosynthesizing bacteria existed that made use of the dead upper layers as a sort of sunscreen, filtering out the dangerous

UV rays and allowing the less energetic visible light to work its wonders. During this time, life also adapted mechanisms to repair the damage caused by radiation in the primordial cells. These same mechanisms may be later used by cells to allow a wonderful new mode of reproduction, the beginnings of sex. But for the moment, the Earth, and this history, is rated G.

The greatest impact of life on planet Earth was yet to come, but our oxygen atom sat out its beginnings. Once again, it got caught up in the bowels of the Earth. Bound up in a complex carbohydrate, its host dead, it became embedded in a lattice of minerals carried by surrounding water. Over time, it was buried, once again, in the Earth's crust. Here it remained for another 200 million years, as the minicontinent it found itself on built up and drifted slowly through the action of plate tectonics, until a collision of plates put this region between a rock and a hard place. Forced downward toward the mantle, the molecule was once again broken apart by the great heat into its constituent gases. Once again the pressure built up, and once again our atom was shot up into the sky as carbon dioxide.

During its 200-million-year sojourn, however, life took its most dynamic risk yet, with implications for the entire planet. In order to complete the photosynthetic process, hydrogen atoms must continue to be harvested to create organic materials and as a source of electrons for energy transfers. Hydrogen sulfide and occasional free hydrogen gas near hydrothermal vents or volcanoes are but bit players in the hydrogen budget of this planet. By far the biggest source of hydrogen on Earth is all around us, covering three-fourths of the Earth's surface, in the oceans of this planet. It was just a matter of time before life, ever opportunistic, discovered this great source of energy.

History often repeats itself. The exigency of existence once drove life, via natural selection, to seek out the hydrogen in the Earth's oceans. These same basic energy needs, only now arising from the machines that help drive human civilization, will eventually force intelligent humans to turn to the oceans as well. Hydrogen is not merely a source for photosynthetic energy conversion of light into matter. It is, via the same process that powers the sun and stars, fusion, the ultimate source of energy. Moreover, the chief products of fusion, primarily helium, are stable, and do not contribute greenhouse gases to the environment. Once we can control fusion on Earth, people of the future may wonder why we bothered with fossil fuels, created by the energy stored up by the same ancestors that first discovered how to break apart water into hydrogen and oxygen.

Photosynthesis, as it is carried out today, really takes place in three stages, beyond the simple light and dark reactions of the anaerobic sulfur bacteria. These two sets of reactions are combined as part of what is now called *Photosystem I* (PS I) reactions. In these reactions, the energy of light is used not only to excite electrons, but to split apart hydrogen sources like hydrogen sulfide, in order to obtain hydrogen atoms for organic use. The problem with using this mechanism to obtain hydrogen from water is that hydrogen and oxygen are very strongly bound in water, and it simply takes more energy to break them apart.

For this purpose, a new kind of chlorophyll molecule is exploited, one that allows a new site for energy production, called *Photosystem II* (PS II), in addition to the standard PS I pathway. At these new sites, light of a slightly different wavelength, one carrying more energy, can be absorbed, helping break apart water molecules and also creating energetic electrons, which are then fed into the PS I reaction network. This extra energy and electron input means systems that utilize both networks can produce more ATP than organisms using PS I alone.

It is likely that this second pathway evolved after systems had perfected the first, early form of photosynthesis. It is doubtful that the biochemically more complex system could have evolved first, especially when the early environment was particularly conducive to the PS I reaction.

Probably the first bacteria to evolve this new talent were ancestors of the modern blue-green cyanobacteria that live atop present-day stromatolites. Indeed, one type of living cyanobacteria has the ability to utilize only PS I when living in high concentrations of hydrogen sulfide, whereas it will revert to the more efficient use of both mechanisms and water as a source of hydrogen when in an appropriate environment. This suggests that these cyanobacteria evolved from earlier sulfur bacteria, with the evolutionary advantage of being able to move out to the open ocean for their source of hydrogen.

Once water itself became a source of life, the whole world became available for colonization by bacteria . . . but at a price. Breaking up water produced the much-needed hydrogen atoms that could provide electrons for energy, and hydrogen as a building block. But at the same time, it produced perhaps the biggest potential threat to life on Earth at that time: pure oxygen gas.

Oxygen, as I have already described, spells trouble for organic metabolism. The root of the problem is simple: Oxygen grabs electrons with a vengeance. Most important, it grabs electrons before they can be used to provide energy to living systems, and in the process it also oxidizes substances such as hydrogen before they can be incorporated into complex molecules. If free oxygen had existed at the beginning of time on Earth, it is likely that no life would have had the luxury of evolving. Almost all the species that were vital to the early development of life on Earth quickly die in the presence of oxygen. Fermenting bacteria and their sulfur-eating descendants all avoid its presence like a plague, even today. They are found only in environments where oxygen is scarce.

Moreover, oxygen quickly destroys the habitat for such bacteria. Hydrogen sulfide, for example, is stable only in the absence of oxygen. Today, hydrogen sulfide that is produced deep under water combines with oxygen before it reaches the surface. For this reason, sulfur bacteria, even if they could tolerate oxygen in their internal metabolisms, are generally relegated to regions near hydrothermal vents, hot springs, or deep underground, where reduced hydrogen is present. These microbes have essentially sequestered themselves in locations that point back to the conditions of the early Earth. Both metaphorically and sometimes literally, they have let the sands of time bury them.

By refusing to adapt to changing circumstances, the hyperthermophilic anaerobic bacteria relinquished their dominance over the Earth, but they provided two services that are invaluable to (at least some of) their modern descendants. First, they offer scientists living pointers to the past. They have retreated to those locations that still resemble those of their primeval beginnings. (Indeed, bacteria that extract atmospheric nitrogen for use in organic molecules even today are limited to mostly anaerobic environments.) Second, more important for the rest of humanity and indeed for the rest of life on Earth, by hanging on to inaccessible locations they insulated themselves against the inevitable

catastrophes that were to follow. Protected from meteorite bombardments, climate fluctuations, and other disasters, they very probably served
to keep the planet alive through the tough times so that one day we
might arise to wonder how.

Even as primordial microbes began to spread to more exposed locales
throughout the oceans, they established protective mechanisms to help
assure their survival. Some species of sulfur bacteria and cyanobacteria
that continue to live in water today incorporate gas vacuoles, like sub-
marine ballast tanks, that can fill with gas or release it to keep them appropriately below the water's surface, where dangers such as ultraviolet
radiation or oxygen would once have lurked.

Whatever the protection, however, for those systems alive at the time
our atom once again became imprisoned in the crust, the photosynthetic production of oxygen by cyanobacteria was in principle a catastrophe comparable to a major meteorite impact. This danger was
perhaps greatest for the primeval cyanobacteria themselves. These
would be poisoned by their own waste products. Even today, there exist
species of cyanobacteria that are intolerant of the oxygen they themselves produce, and must live in close proximity to microbes that take up
the oxygen as soon as it is made.

But there were no such microbes 3.5 billion years ago. Had oxygen
built up quickly in the atmosphere, it would have spelled the end
for most of the life then on Earth. It might seem inexplicable that small
single-celled creatures, merely thousandths of a centimeter in size,
could have so powerful an effect on the Earth's environment. But as
H. G. Wells noted in *War of the Worlds*, even microscopic microbes can
pack quite a wallop if you are unprepared for them. Working in concert,
trillions and trillions of cyanobacteria would eventually reshape the
atmosphere of the Earth. Thankfully, the Earth cooperated, so that it
would take more than 1 billion years for the patient cyanobacteria to
make a significant dent in the biosphere. This would turn out to be ample time for life not only to evolve efficient new protection mechanisms,
but to ultimately turn this new liability into an asset that would one day
make life as we know it possible.

The reason that this buildup is slow is straightforward. Oxygen is so reactive, it would quickly react with anything it encountered in the envi-

ronment. So many oxygen sinks existed on Earth 3.5 billion years ago that even trillions of cyanobacteria puffing out bursts of oxygen for 1 billion years on the surface of the Earth were about as effective as the wolf huffing and puffing at the brick house of the third little pig.

After all, the Earth was in a largely reduced state. The entire planet was waiting to be oxidized. Hydrogen sulfide, reduced carbon in organic materials, and hydrogen gas itself were all waiting to suck up oxygen. Even the rocks were starving for oxygen, taking it up as calcium carbonate in limestone, or through oxygen-greedy iron, uranium, or sulfur.

Indeed, iron had long served as a protector of life against oxygen, in several different ways. Ultraviolet radiation in the early atmosphere would periodically break apart water vapor into hydrogen and oxygen. Iron ions in a reduced state can dissolve in water, and the early oceans must have contained substantial dissolved iron from erosion off of the early continents, and from the mantle upflow in hydrothermal vents. Oxygen in the atmosphere can dissolve in water, and will oxidize the iron therein. Once iron has been so oxidized, however, it is no longer soluble in water, and it precipitates out into a reddish solid that falls to the sea floor. The amount of such ferric compounds, as they are called, found in early rocks, from about 2.5 billion to 3.5 billion years old, when there were few large sedimentary basins, would have required simply the amount of oxygen that would naturally have been created by radiation in the atmosphere. However, from about 2.5 billion to 1.8 billion years ago, huge deposits of what have become known as *banded iron formations* exist. Extending over ranges up to 1,000 kilometers long, and up to 1 kilometer thick, these could not have been formed by the intake of oxygen produced by radiation alone. The fact that they are banded, with red layers interspersed with other layers, presumably indicates that the oxygen abundance at this time was still variable.

The fact that these deposits fall off after about 1.8 billion years ago tells us that almost all of the unoxidized iron available in the oceans had been exhausted. The likelihood of this buildup of oxygen around this time is supported by several other pieces of evidence, one involving the iron compound pyrite, which may have been so important to the formation of the earliest life. Pyrite is easily oxidized. Nevertheless, deposits of pyrite dating back 2 billion to 3 billion years are found in regions where streams must once have been flowing. Had oxygen been around at the time, the pyrite would not have survived.

Early fossil soils containing iron also help provide a timeline for oxygen buildup in the atmosphere. In modern soils, any iron leached from rocks is quickly oxidized, so iron accumulates near the surface of the soil. In ancient soils, however, with little oxygen and a great deal of carbon dioxide in the atmosphere, the iron remained soluble in water, so it percolated downward through the soil, and the iron is thus concentrated near the bottom of the soil column.

Finally, uraninite, a compound of uranium, gives a date for the onset of substantial oxygen in the atmosphere consistent with that obtained from iron, and also from sulfur deposits, using a kind of opposite process to that involved with iron. For uranium, it is the unoxidized form, uraninite, that is insoluble, while the oxidized form can combine with carbon dioxide to dissolve in water. Measurements of uraninite survival suggest that by about 2.5 billion years ago, oxygen in the atmosphere was at most 0.3 percent of its present abundance (still high enough so that ozone created in the upper atmosphere was sufficient to absorb ultraviolet radiation and thus protect life). At the same time, the carbon dioxide level had already been substantially reduced, if not yet to its present level, to at most a few hundred (indeed, one study suggests at most 30) times that value. It is interesting that around this period, the formation of continental crust increased dramatically, as global plate tectonics, and the associated extraction of carbon dioxide from the atmosphere, had begun in earnest.

Undaunted, the microscopic cyanobacteria kept pushing out oxygen, day in, day out, for years, centuries, millennia, eons. Free to use water, they began to occupy the oceans. Yet up until about 2.5 billion years ago, sinks for any photosynthetically produced oxygen effectively kept the oxygen threat in check. Sometime around this period, and over the next half billion years or so, the oxygen abundance in the atmosphere built up to perhaps 10 percent of its present value. Microscopically small creatures had managed to make up for their small bulk by large numbers and persistence, and in so doing, they first colonized the world, then they changed it. In the intervening billion years, other forms of life developed that could thrive in this new, oxygenated world.

❋

When our oxygen atom re-entered the biosphere, perhaps 3 billion years ago, the differences in the new above-ground world were apparent in ways that would have directly affected its life cycle. The land was slowly turning green with chlorophyll. Photosynthesis had already begun, perhaps even oxygen-producing photosynthesis. Thus, albeit for short periods, our oxygen atom, when entering life's metabolism in the form of water, could have been freed by photosynthesis. It would have then existed as pure oxygen gas, perhaps for the first time since its interstellar travels following its creation in its parent supernova. These periods of freedom would have likely been brief, alas, as everywhere were rocks and organic materials waiting to be oxidized. At times our atom might have made it back to carbon dioxide, or water, where it might have survived for tens or hundreds of millions of years. But most likely it was once again precipitated out on the ocean floor, either as part of a carbonate rock, an oxide of iron or sulfur, an organic molecule, or perhaps as part of a dead cell, again to be subducted and regurgitated much later by the Earth.

There may be some romance in the realization that the oxygen we are breathing in right now was first breathed out (if you can call it that) in its pristine form by a single-celled organism clinging to a rock in this primeval world. But once you calm down and get over the rush, things are still rather staid as far as the oxygen is concerned. The ultimate oxygen cycle is yet to occur. At this point, oxygen remains essentially peripheral to the whole life process — other than existing as a veiled menace to be avoided. When it is part of the life cycle, it is carried along merely for the ride. Before the invention of PS II, oxygen entered living metabolism as part of carbon dioxide, which was broken up primarily for the carbon. Our oxygen atom was either incorporated as part of an organic molecule, perhaps accompanying a phosphate on an ATP chain, or excreted as water.

Even in the new photosynthetic process PS II, oxygen is produced largely as an irrelevant by-product in a chemical reaction that is both violent and uncaring. Oxygen enters as a part of a water molecule which is rudely broken up for the valuable hydrogen. Two hydrogen ions (that is, protons) and their accompanying electrons are dragged away from the water molecule like children from their mother. They are then temporarily enslaved in work camps, as the electrons, powered with energy absorbed from light, move in a chain gang to service the energy needs of the organism. They pump and prod along their proton cousins, which in turn help create ATP before perhaps being allowed to leave the system with other OH parents as part of water, or perhaps to be permanently assimilated into the ever-growing organic complex.

The original oxygen is simply excreted as waste, in the form of oxygen gas, or, if it pairs up with some waste protons, it leaves once again as water. The oxygen gas will alter the external environment, to be sure, but the oxygen itself has still not yet entered the cycle of life in any direct or vibrant way, in spite of the great potential it offers. Perhaps it was lucky. Up to this point it was not an exploited worker. I suppose it depends on your point of view. Exploitation, or useful work? In any case, like many an immigrant, oxygen was still shunned by much of life in the early archaean era. But as the waste-oxygen abundance built up over these eons, life soon adapted to cope with, and ultimately exploit, this exotic and dangerously reactive partner. Afterward oxygen would never have a free ride again.

Hold your breath while you are reading this page. After the first paragraph or so, depending on your reading speed and lung capacity, of course, a certain urgency begins to take hold, doesn't it? By the time you have finished a full page, depending on your lung capacity, you will at least have begun to feel slightly uncomfortable. What begins as a slight strain develops into an irresistible urge to breathe, and you may also begin to feel a pain in your chest. Your head may even begin to throb. If you have the discipline and the desire to read on long enough, you will begin to feel dizzy. Eventually, if your body would let you continue to hold your breath, you might black out. But do start breathing again. The rest of the story is too good to miss.

We can do without food for days, even weeks. The need for water is more urgent, and dehydration will usually kill before starvation will. But the need for oxygen is the most powerful need of all. Without it, we are all dead within the course of not days or hours, but minutes.

This fact is all the more surprising when you realize that the ultimate role of oxygen in our bodies, a role that magnified life's energy-producing capability by perhaps an order of magnitude, is simply to pick up waste electrons at the end of their wild ride through our metabolism. But without this pickup routine, where would we be? Life requires energy, and complex life requires energy beyond all expectation. Each time useful work is carried out, waste heat must also be produced. It is an inexorable law of physics. All this work takes energy, lots of energy.

Here is a statistic that I find amazing. *The average male human uses about 420 pounds of ATP each day of his life to power his activities.* Considering that we all contain less than about 50 grams of ATP in our bodies at any one time, that involves a lot of recycling. Specifically, each molecule of ATP must be re-energized at least 4,000 times each day.

The respiration of oxygen, even though it is tied to the very last stage of energy production in our bodies, increases the energy output obtained from a sugar molecule from 2 molecules of ATP purely via the process of fermentation, to 38 molecules of ATP when respiration is added. Without it we simply lose power and, like the Eveready Bunny's rival, we quickly stop working.

The process of respiration begins very similarly to fermentation. Glucose is first broken down to acids such as citric acid. At this point, however, a new cycle takes over in which high-energy electrons are deposited on carrier molecules that then take them to an electron transport chain that produces up to 32 additional ATP molecules. Finally, at the end of this chain, the weary and almost completely de-energized electrons and their accompanying protons are simply grabbed by oxygen, which is then reduced to form water.

That is the whole story of the powerhouse behind modern life, and oxygen comes in only at the very end. It looks so simple that it hardly seems worth a few hundred million years of evolution.

If things are so simple, then what is the big deal? Why did early life avoid oxygen at all costs? The point is that getting oxygen to suck up electrons is no problem at all. The problem lies in stopping them from sucking them up too soon. Left to its own devices, oxygen would take up electrons before any ATP could be produced at all, undercutting the whole metabolic pathway of life.

Remember that respiration is *controlled* combustion. Fire is uncontrolled combustion. The difference is clear even to the untrained observer. Assuring that oxygen could delay its gratification required the development of an army of biological machinery, including enzymes and proteins, along with the RNA and DNA that would encode the recipes for their formation. This is why respiration is only for the experts.

We can see a hint of the sophistication required to handle oxygen when we consider the history of the hemoglobin molecule, a molecule that surprising new discoveries are being made about even today. Hemoglobin is the molecule that carries oxygen through the blood to the sites where it is effective in accepting electrons. Without hemoglobin, blood can dissolve only about 1/70 the amount of oxygen it can carry with this molecule present. The workhorse in hemoglobin comprises 4 porphyrin rings, similar to the one used in primordial electron transport chains. At the center of each ring is an iron atom, making it a heme group, which can alternately bind and then release oxygen, just as the heme group of porphyrin allows that molecule to pick up and drop off electrons when needed. More important, perhaps, the oxygen is not set free where it might do harm.

Hemoglobin is essential to the exchange of gases within our bodies. Even minute changes in its internal charge can drastically affect its ability to transport oxygen, as Linus Pauling discovered when he investigated the hemoglobin of humans and its relation to sickle cell anemia. Yet the oxygen transport and drop-off character of human hemoglobin might hint at a far more important evolutionary solution to the early vexing problem of oxygen. For example, a form of hemoglobin that binds very strongly to oxygen is found in a common anaerobic parasitic worm in humans and mammals. The hemoglobin binds to oxygen in order to protect the worm from any free oxygen that may be around. Could it be that the precursors to modern hemoglobin molecules originally evolved not to transport oxygen into the body, but rather to isolate errant oxygen atoms that might stray too close to organic reaction centers? In this way they might have protected early species from the potentially lethal appetite of oxygen for electrons.

Some completely unrelated evidence of the possible historical protective role of hemoglobin comes from another kind of worm, found as far away from mammals as one can get on Earth. This worm is one of the weird yet relatively advanced life-forms that thrive today near hydrothermal vents at the bottom of the oceans, where earlier, archaean life undoubtedly thrived. Shortly after these black smokers were discovered, huge colonies of long red wormlike creatures, now called giant tube worms, were found to flourish nearby. These worms are up to 1.5 meters in length and 40 centimeters in diameter.

When these worms were discovered in these hot, acidic, and sulfurous environments, biologists were naturally eager to determine how they could grow to such large sizes in such hellholes. It turns out that these animals contain inside them bacteria that can combine oxygen with hydrogen sulfide, and use the energy released to build organic carbon compounds from carbon dioxide — a modern aerobic version of the anaerobic sulfide oxidation into pyrite that may have been associated with the first spark of life on Earth. There is a problem, however. Hydrogen sulfide is generally toxic to aerobic animals like the tube worms. Here is where the hemoglobin in the tube worm blood comes into play. It has a very high affinity for sulfides, and presumably can transport these, as well as oxygen and carbon dioxide, down to the bacteria deep in its interior, while protecting the rest of the animal from sulfide poisoning. At the same time, the high oxygen binding of the hemoglobin protects the bacteria from too high an exposure to oxygen, and presumably it also makes certain that the sulfides in the blood are not oxidized before they get where they can be used. The large tube of the tube worm allows it to scavenge the waters for these gases, where the bottom vent water mixes with the ambient seawater. It picks up carbon dioxide and sulfides from the former, and oxygen from the latter. Hemoglobin and other proteins then allow these potentially toxic gases to be held at bay, for later use by the symbiotic bacteria in the worm's lower regions.

Because of the long history of hyperthermophilic life near hydrothermal vents, it is natural to wonder whether the protective role of the hemoglobin and proteins in tube worms might hark back to an early form of protection for their anaerobic ancestors as the first onslaught of oxygen began, some 2.5 billion to 3 billion years earlier.

In any case, life did slowly evolve the capacity to regulate and use oxygen, and this changed everything. Suddenly, energy usage heretofore unthinkable was now possible, and the world of life could diversify and grow in ways that previously were physically impossible. Each new process of life generates waste heat along with usable energy. Bigger systems, with more processes, generate more heat. Human beings, for example, even when resting, generate almost 100 watts of heat on a continual basis. This is why a crowded auditorium gets hot. A room with 100 people in it has the equivalent of a 10,000-watt heater! Only through oxygen respiration can this kind of energy be generated as waste heat by biological systems. Only then can life be ready for the big time.

Of course, it would take on the order of a billion years for oxygen levels to rise to anywhere near their present level and, as we shall see, even longer for life to fully exploit this potential. For a billion years, cyanobacteria and stromatolites ruled the world, growing in larger and larger clusters, sometimes 100 meters high, because there was virtually nothing else around to threaten them. They diversified, moving into every location imaginable, from the ocean to lakes to deserts to the inside of rocks. As the oxygen abundance rose, more species of bacteria learned to cope with this danger, and then exploit it. Many different species of oxygen-respiring prokaryotes began to develop. With the new energy, larger cells could function, and even multicellular animals began to form.

As far as our oxygen atom is concerned, however, the new phenomenon of respiration brought about a more immediate and dramatic change in its life cycle. As the oxygen abundance in the atmosphere increased, once the oxygen sinks like iron had all been oxidized there was less likelihood that our oxygen atom would be prematurely stolen from the atmosphere. It could now be recycled while participating vitally in the processes that govern life. By simply accepting tired electrons at the end of a long work cycle, gaseous oxygen could be converted to water in slow-motion combustion. As a part of water, the oxygen could once again be liberated by photosynthetic plants that harness the light of the sun. Or, as part of carbon dioxide, these same processes could convert it to water. Now oxygen could be continuously and rapidly transformed from carbon dioxide to water to oxygen and back again, on a timescale of hours or days, not millions of years.

Indeed, the first bacteria to produce oxygen were also probably the first ones to evolve an ability to use it. Cyanobacteria photosynthesize during the day and respire at night. Their machinery is complex enough to handle either process, but not at the same time. Later on, other algae and plants would develop specialized components that could handle different processes simultaneously, but the simple undifferentiated cells of the cyanobacteria were not yet up to the task. Nevertheless, the cyanobacteria's strategy has a certain logic and efficiency. During the day, light energy is exploited, and at night, dark energy. From these bacterial-mat-covered outcroppings of land, waves of oxygen would rise when the sun was shining, and carbon dioxide would rise toward the moon at night.

This recycling also means that our oxygen atom has a far richer history, even during these primeval times, than it would otherwise have had. It was freed into the atmosphere by a cyanobacterium more than 3 billion years ago, and over the eons to follow was exhaled and reabsorbed in one form or another by a veritable host of slimy creatures. It was part of the slime, part of the green mat, free in the evolving air, part of the rocks, and part of the water. Over the course of a billion years, it explored every nook and cranny of our planet, from high above the surface to the bottom of the sea, and every now and then, even the bowels of the Earth.

This newfound energy allowed life to truly begin to change the face of the Earth. Biochemistry, combined with urgent geological forces, had a devastatingly powerful impact. Continents were forming and growing, carbon dioxide was being extracted from the atmosphere and fixed in the ground, and life was adding oxygen slowly and steadily to the mix. The rocky surface of our blue planet was now getting covered with green. The oceans, covering three-fourths of the available light-gathering area of the Earth, were being colonized as well. Hordes of plankton, located 80 meters or so below the water surface to protect them from ultraviolet radiation, were churning out oxygen, and still dominate the production of oxygen and organic materials in tropical waters today. And even the skies were becoming safer. As the oxygen levels increased, ozone was produced by interactions with solar radiation, and this built up a protective sunscreen for the planet, absorbing ultraviolet radiation before it could harm life below.

The oxygen level on Earth never rose very much above its present level of about 20 percent of the composition of the atmosphere, which is a good thing. Oxygen is so reactive that spontaneous uncontrolled combustion begins if the atmospheric fraction grows too high. The spark of life, in this case, could literally set off a conflagration. Even damp plants are very flammable under high-oxygen conditions. Indeed, some types of coal from the later, Carboniferous era of giant insects when oxygen levels may have been higher, provide evidence of intense local forest fires at that time. One again senses that there must be some feedback mechanism which raises the oxygen abundance so life can exploit it, but fortunately limits it below the danger point.

During this blissful period, life could truly enjoy the fruits of its labor, and the planet reached a new equilibrium. The sacs that made up prokaryote cells were simply too primitive to allow truly complex organisms to evolve, however. To take the next step, the basic building blocks of life would themselves have to evolve.

In 1967, Lynn Margulis helped pioneer a sensible idea. We have seen that single-purpose microbes can have symbiotic relationships with each other, as in stromatolites, and with more complex life-forms, as in tube worms. Margulis explored a possibility that had been around for a while, namely, that the complex cells of advanced animals and plants, ones with a nucleus and many separate organelles within them, each with its own purpose, could form by certain bacteria cells simply assimilating other special-purpose bacteria within them. They would then employ their new drone's special talents for their own purposes, as the Borg do on *Star Trek*. In this way eukaryotes were born, organisms whose cell(s) had a nucleus and separate components, each with a specific task.

Such a possibility is certainly reasonable. Cells might host photosynthesizing bacteria containing chlorophyll, or other bacteria that effectively respire oxygen. Let's focus on the latter case for a moment. Such bacteria could form a symbiotic relationship with an anaerobic cell, perhaps invading it to feed off the organic waste of its host. Eventually, as oxygen built up in the atmosphere, these cells obtained an evolutionary advantage, as the parasites within them could process oxygen to produce energetic compounds. With their protective nuclear enclosures, and their talented symbionts, these cells were not restricted to the increasingly rare anaerobic environments on Earth. Eventually, life inside these cells became so easy that the symbiont bacteria no longer needed a life outside, and began to shed unnecessary genetic material.

The proof of the pudding would be if one could show that the individual components of modern cells, in this case objects like chloroplasts and mitochondria, which respectively have the functions mentioned above, bear more of a resemblance genetically to certain free-standing prokaryote cells than they do to each other. Studies of the makeup and antibiotic sensitivities of the ribosomal RNA of mitochondria and that of several bacteria, for example, suggest this is in fact the case.

This new level of complexity was probably necessary before sophisticated animals and plants could form on Earth. In any case, it was a vital development on the road to the modern era. But while such cell cannibalization is very important biologically, from our point of view, which is also the point of view of our oxygen atom, it is just more of the same. Whether the oxygen would be incorporated in a simple bacterium or in an organelle of a complex cell or transported between such cells makes little difference. The end result is the same. Life itself was becoming more diverse, but the cycle of life, as far as oxygen is concerned, remained largely unchanged.

Nevertheless, new creatures could take far better advantage of the new energy allowed by oxygen burning, and complex multicelled plants and animals eventually came into their own following the increase in oxygen in the atmosphere about 1.5 to 2 billion years ago. The beasts that lived during this era were like the vision of the Ancient Mariner: slimy creatures with and without extremities. Diversity began to develop and, on the surface, the planet began to appear much more like the planet we live on today. Oxygen levels approached their modern value, plants lived in soils on the land and in the sea, and animals swam in the oceans. Yet something was missing. With all this great new potential, life seemed to be in a rut. So great a rut, in fact, that for hundreds of years, dedicated paleontologists could find no evidence that it ever existed at all during that 1,000-million-year period.

Perhaps a certain complacency set in. Was it simply that life was too easy? Did life need a new niche in order to develop? If it did, nature would soon provide one. For the future of life was not, and is not, purely in its own hands. Circumstances beyond its control would govern the incredible changes that were to follow — perhaps the biggest revolution ever to occur in Earth's history following the origin of life itself — and our oxygen atom, like all of the other atoms on the surface of the Earth, would be carried along for the ride.

Dramatic and powerful physical processes would determine the nature of the planet our oxygen atom, and we, inhabit today, just as they would shape precisely how the biochemical revolution that carried life past the threshold from ancient to modern would be exploited.

PART
THREE
RETURN

The future ain't what it used to be.

YOGI BERRA

15.

A SNOWBALL IN HELL, HUMANS, AND OTHER CATASTROPHES

This too shall pass.

SUFI MANUSCRIPT

L ord tunderin' Jesus! What are ya at? How's she goin', buddy?"

As far as I know, there is only one location in the world where you might receive such a greeting. That is the island of Newfoundland. There is no other place like it, and that is as it should be. Generations of lives spent either on the sea or in isolated hamlets on the coast waiting to go to sea have created a special language, and a special sense of warmth and humor.

I grew up in Upper Canada, as those from the eastern provinces like to call it, where I was schooled in "Newfie" jokes, which were safe to tell because none of us actually knew any Newfies. But thanks to my wife, who hails from Newfoundland's neighbor Nova Scotia, I have since learned to appreciate the depth of humanity that seems to overcome the poverty and loneliness that can go along with a sea-based economy. Natives of Nova Scotia and New Brunswick, like their Scottish and Irish ancestors, are professional storytellers. The true masters among them come from the northeastern corner of Nova Scotia, Cape Breton Island, which is separated by a very narrow strait from the rest of the province. Here Celtic roots combine with traces left from the Acadians, the French populace who were forced out of this area by the British over 200 years ago.

Some of the expelled Acadians later settled in Louisiana, where the word *Acadian* evolved into the term *cajun*.

But the Cape Bretonners are tame when compared to their neighbors from Newfoundland, at least when it comes to displaying a colorful sense of language, and the ability to party through the evening. Anyone from Newfoundland would be happy to tell you a story explaining why his or her home is unique, but in 1990 paleontologists discovered an unassailable reason why Newfoundland is special. For it is here that the modern world, as it is now officially defined, began.

For a billion years following the first appearance of multicellular eukaryotes, life on Earth continued to diversify, apparently without major incident. At the same time, primitive bacterial mats continued to flourish, because the new organisms that came into existence didn't seem to feel the need to graze upon them. During this time, oxygen was still building up relentlessly in the atmosphere, and new and larger life-forms were developing to take advantage of it. The cell nuclei that define eukaryotes harbor the sensitive genetic information needed for replication, but the picture presented in the last chapter of complex cells arising by accreting special-purpose bacteria suggests that oxygen had an evolutionary role to play in this development as well. The segregation of material inside a nuclear shell meant that it was protected from the harmful effects of the oxygen that was nevertheless needed to power the functions of these cells.

This meant that our oxygen atom was chaperoned throughout its time in these living systems, and visited only certain special locations. Recall that mitochondria, for example, are structures inside these new complex cells that govern their respiratory function, and it is here that oxygen helps power the production of ATP by removing the used electrons. Before nucleated cells came into existence, our oxygen atom was free to roam over the entire cell. After that time, it was forbidden in the inner sanctum unless already a part of a larger complex organic molecule. Eventually, the mitochondria became the only place free oxygen was welcome. Similarly, as eukaryote cells accreted more components, the specific paths followed by oxygen inside living species became more diverse, along with the growing diversity of life itself.

The accretion and assimilation of life-forms inside cells in order to help them function was followed by the development of the first true animals. These creatures live by utilizing the photosynthetic work done by other species. But they do so not by incorporating these species within their cells, but rather by digesting them and breaking them down into their energy-rich components. And they require oxygen to efficiently burn these components. Animals can exist only because of the hard work done over the ages by photosynthesizers, first in producing organic materials to eat, and second in producing sufficient oxygen to fill the atmosphere for use in respiration. Human are simply one in a long line of animals exploiting the fruits of the labor of plants and bacteria over the eons.

By 600 million years ago, the surface of the planet would, in principle, have seemed familiar, and even hospitable, to a time traveler from the present. Green plants would have sprouted from the soil, and wiggly animals would have been found swimming in the seas. Indeed, except perhaps for possibly poisonous plants or fungi, few threats would have lurked for our explorers. No Jabberwocks, with jaws that bite or claws that catch, hid behind the rocks.

And it is precisely this absence that is so noticeable. Until almost 600 million years ago, diversification of life on the planet had led nowhere particularly special. Multicellular animals were only just arriving on the scene, and the skeletons that would eventually make possible the great animals, from dinosaurs to blue whales, from birds to humans, had yet to appear.

Then, suddenly and quickly, everything changed, everywhere on Earth. The Cambrian revolution had begun.

The changes are observable on the walls of the cliffs of southern Newfoundland, near a place with a typical Newfie name, Mistaken Point. They are also visible in sedimentary deposits at various locations around the world, from the Ediacara Hills of southern Australia, where Precambrian animal fossils were first observed in 1946, to China, to the Siberian Arctic, to England and Wales. Mysteriously and within only a few million years, life shifted direction everywhere on the planet.

For the first time, many species seemed to disappear. Instead of diversity increasing and encompassing that which had gone before, life appeared to qualitatively change. Small animals with shells began to

appear. These shells would persist as fossils, which tricked modern scientists into believing that life itself began at this time. Instead, we now
know that life had merely crossed a great threshold into the future. With
shells and skeletons, the groundwork was laid to build bigger beasts. Just
as big bridges and big buildings require a rigid infrastructure, the laws of
physics require large land animals to have something firm holding them
together.

Would this threshold have been noticeable in the life of our oxygen
atom? Clearly yes, if the oxygen happened to get incorporated into one
of these shells or skeletons, and was preserved in a rock for the ages. But
what is really relevant here is whether a "generic" oxygen atom would
notice the onset of the Cambrian era. This would only be the case if the
change from Precambrian to Cambrian was associated with some other
large-scale physical or chemical change on Earth. The extinction of
some species at the Cambrian–Precambrian boundary is suggestive, but
for some time the biological explosion associated with the Cambrian era
was thought to have been essentially purely biological in origin. It had
been recognized that shortly before this time, the existing continental
crust formed a supercontinent strung out along the equator, and this
landmass had just begun to break up. This, it was noted, would have dramatically increased the total amount of coastline, since many smaller
continents, each surrounded by water, would have more coast than a
single large continent. With more coast near the equator come more
warm and shallow tidal areas. Since tide pools were thought to be good
breeding grounds for new life, it was reasoned that their appearance
might explain the sudden Cambrian explosion. But more recent discoveries, made over the past decade, suggest a dramatically different state of
affairs. At the time that the diversity of life was about to explode, it appears that the Earth froze like a giant snowball.

From the point of view of a physicist, which is, I think, not too different from the point of view of our oxygen atom, catastrophes such as this
break up what would otherwise be a pretty monotonous progression.
The remarkable diversity of life, and the mechanisms for survival, pro-
creation, social patterns, the origin of consciousness, and so on, are un-
deniably fascinating in their own right. Yet once the basic mechanisms
of photosynthesis and respiration were established, and nucleated cells

originated with their various organelles, the future life cycle for our oxygen atom was largely just more of the same. Biological evolution can progress by exploiting every different combination of the basic biochemical units, and those that are evolutionarily favored survive and procreate. As far as oxygen is concerned, however, the routine is more or less the same: Free oxygen oxidizes organic materials, burning them to carbon dioxide and water. In the form of carbon dioxide, oxygen can either get bound up in organic molecules, sediment out in carbonated rocks, or via photosynthesis be converted to water and then returned as oxygen to the atmosphere . . . and so on. Whether the process occurs inside a stromatolite or a shark is merely a minor perturbation. More of the same — except when it isn't! When something dramatic changes the ground rules, then that is worth writing home about.

This new episode in the lives of our atom, proposed in 1998 by the geologist Paul Hoffman and his geochemist–oceanographer colleague Daniel Schrag, begins when the supercontinent of Rodinia first began to break apart around 750 million years ago. As the smaller subcontinents drifted apart, oceanfront real estate became far more accessible. Moreover, this real estate was primarily tropical, since the continents were clustered close to the equator at this time. Increased sources of moisture in these tropical regions brought more rain. With more rain, more carbon dioxide was scrubbed from the atmosphere, and global temperatures begin to fall.

The sun at this time was still somewhat less luminous than at present, but the greenhouse effect had managed to keep the Earth from freezing over even at earlier times, when the sun was even less luminous. Normally as the Earth cools, ice forms over the continents (once continents exist). This ice provides a barrier to the formation of carbonate rocks from carbon dioxide in the atmosphere, allowing volcanic sources of carbon dioxide to cause it to build up in the atmosphere once again. This maintains the greenhouse effect, and keeps the Earth warm.

With the continents located near the equator, however, ice did not build up on the rocks as the global temperature fell, allowing carbon dioxide to continue to be scrubbed from the atmosphere. At the same time, ice was building up over the rest of the Earth. Being white, it reflected a far greater proportion of the sun's radiation than did liquid water. With more light reflected, and less absorbed, the Earth cooled even more. The combination of falling carbon dioxide rates and greater reflectivity of the Earth became critical once the ice buildup reached down as far as 30 degrees north or south of the equator (the former is roughly the latitude of Orlando, Florida), and a runaway freezing began to occur. Within 1,000 years of this first large-scale ice build-up, the entire Earth froze over.

On hearing this, one becomes suspicious. First, if the Earth froze over, what happened to life? Second, once it froze, what caused it to thaw? These very suspicions caused scientists to doubt for a long time that such a global deep freeze could have occurred after life had started to evolve into the forms that left the relatively continuous fossil record of the past 3.5 billion years.

With the recognition that life could exist, and indeed thrive, in extreme environments, this first concern began to evaporate. Heat escaping from hydrothermal vents would stop the oceans from freezing through to the bottom. Organisms such as sulfur bacteria could then easily survive under the global ice cover. Moreover, some cyanobacteria today survive even in icy habitats, as can other species. And certainly hyperthermophiles could thrive underwater through the hard times.

The second concern is even easier to allay. As I indicated earlier, volcanoes can replenish the present carbon dioxide content of the atmosphere in less than 1 million years. With all of the continents covered in ice, there would be no sink for carbon dioxide, so that volcanic activity would continue to raise the carbon dioxide level by a factor of 1,000 over perhaps 10 million years, causing a massive greenhouse effect once again. Once carbon dioxide built up to more than 350 times its present-day concentration, this would cause massive melting of ice, especially near the equator. As the seawater evaporated, water vapor combined with carbon dioxide to drive the greenhouse effect further. Global surface temperatures jumped from freezing to perhaps 50 degrees Celsius in a few centuries. The Earth proceeded from an ice cube to a torrid tropical hell. For centuries, torrential downpours would quickly scrub the high carbon dioxide concentration out of the atmosphere once again, producing huge carbonate buildups on the ocean floor. The whole process of freezing and thawing might continue once again. Indeed, it is claimed that it may have occurred as many as four different times between 750 million and 580 million years ago.

This picture is not yet universally accepted, but it does have several attractive features, and moreover, it explains otherwise inexplicable data. At the same time, the global nature of the subsequent Cambrian revolution and its speed suggests some global catastrophe may have predated it. Becoming a gigantic Popsicle certainly qualifies as one. However, it is important to note that the last glaciation of Snowball Earth and the Cambrian biological explosion were separated by 40 million years, so any connection between them must be more subtle. In any case, the evidence that first led researchers to the Snowball Earth hypothesis has little to do with biology.

Over many years geologists found evidence on several different continents of a widespread early period of glaciation. Glacial ice makes particular markings as it moves across rock, and these marked rocks had been unearthed in a wide variety of locations, all around the same time. Moreover, when rock is first formed in the magnetic field of the Earth, some of the magnetic materials in the rock become frozen in the direction of the Earth's magnetic field. The fact that the magnetization of the rocks that exhibited evidence of glaciation pointed parallel to the Earth's surface tells us that these rocks hardened near the equator. How such massive glaciation could occur in the most tropical regions of the Earth remained a great mystery.

Compounding this mystery was another one, which related more directly to our oxygen atom. Mixed in with the glacial debris from this era are deposits of iron-rich rock. It is difficult to imagine how this much iron could build up in the sediments, once oxygen had built up in the Earth's atmosphere. Recall that once oxygen is abundant, iron quickly oxidizes and precipitates out of water, forming sediments. The banded iron formations a billion years earlier represented the time when the available iron in the water oxidized in the presence of the increasing oxygen abundance, and were removed. How could another buildup of iron take place in a world full of oxygen?

In 1992 the geobiologist Joseph Kirschvink proposed a solution. If the oceans were covered with ice for millions of years, this would effectively separate the water below from the oxygen above. In this way, iron could build up in the oxygen-deprived oceans. Once the ice melted, and atmospheric oxygen could again mix with the seawater, the iron would quickly precipitate out and mix with the glacial debris.

Another not unrelated effect would be expected to occur following a very quick thaw. With massive amounts of carbon dioxide in the atmosphere, combined with the torrential rains, huge amounts of carbonate rock should have formed very quickly as the rain scrubbed the excess carbon dioxide from the atmosphere. This also resolves another longstanding mystery. On top of the glacial deposits found from this era are large blankets of carbonate rocks, which normally form in warm waters where rain mixed with carbonic acid leaches the rocks. Why such warmwater formations would be created immediately following glaciation was previously puzzling. In addition, some of the crystal carbonate formations (called *cap carbonates*) found in Namibia indicate that they formed very rapidly, from water highly saturated with calcium carbonate. Again, all of this is readily explained if a sudden thaw followed a deep freeze. Estimates of the amount of carbonate material that would form as the hothouse carbon dioxide was scrubbed from the atmosphere suggest enough to cover the entire present continental crust to a depth of 5 meters!

The final bit of evidence is also familiar to us, and has to do with the ratio of carbon-13 to carbon-12 in these carbonate formations. Recall that life prefers to work with carbon-12 rather than carbon-13. Thus the carbon left to form carbonate rocks when the oceans are full of life has an excess of carbon-13 compared to the normal ratio of these elements emitted in volcanoes, for example. The cap carbonates in Namibia show a rapid fall in the carbon-13–to–carbon-12 ratio to approach that from volcanoes just before the glacial deposits, with a subsequent recovery afterward. This is explicable if the abundance of life dropped as the oceans froze over, and built up again after the end of the thaw.

None of this evidence is by itself definitive, but the combination is at the least suggestive, and at its best compelling. The old saying that if it walks like a duck, quacks like a duck, and swims like a duck, it probably is a duck is relevant here. Looking not at ducks but at their ancestors, the biological record is also temptingly consistent with this scenario. The deep freeze would have killed off a number of species. Moreover, hyperthermophiles, at least, would have been well primed to survive in the hellishly hot summer following this several-million-year-long winter. But though it is tempting to imagine that Snowball Earth may have killed off everything but extremophiles, this idea is definitely not supported by the fossil record. And the fact that Snowball Earth and the Cambrian explosion are not precisely coincident makes it clear that any connection of the former with the extinctions and explosion of diversity in the latter is not so direct.

Nevertheless, it is true that rapid and varied environmental stresses are often associated with massive genetic changes. Snowball Earth was a time when the stresses were both extreme and rapid. The later genetic changes from the Ediacaran fauna to the Cambrian animals are indeed extreme, and seem inexplicable without some other dramatic associated events. Finally, the huge diversity that is so characteristic of the Cambrian era, and appeared to be lacking before that, may also be understandable in an era following a global freeze. During such a time, local communities might be isolated from one another. Such isolation has always generated new species on Earth.

As those who make money in a bear market keep claiming, there is always opportunity, even in adversity. Certainly that seems to be the case for the progress of life on Earth. Perhaps without global calamities, successful life is not driven to change and develop. The maxim "If it ain't broke, don't fix it" certainly explains why cyanobacteria have continued to successfully populate the Earth in one form or another for over 3 billion years! In any case, while speculation about the possible generative impact of a Snowball Earth event is, at the present time, merely that, it is virtually certain that without calamities of one form or another, we would not be here today.

And it is the calamities that mark the ages of our oxygen atom from before the Cambrian explosion to the present time. A global snowball phase would have dramatically altered the experience of our oxygen atom on Earth, which is one reason I am particularly partial to it. If our oxygen atom were existing in its free form in the atmosphere prior to the deep freeze, which I shall assume, once the Earth's surface froze over (note that since Hell is supposed to be deep inside the Earth, the scenario I have described would *not* have it freeze over), the opportunity for mixing and evolution would largely cease. Oxygen would persist in the atmosphere, and perhaps build up in abundance somewhat if some photosynthetic surface creatures continued to eke out a livelihood in this otherwise frozen wasteland. But the great cycles between oxygen, water, and carbon dioxide that govern life, climate, and geology on Earth would be altered for a short period which, we should nevertheless remind ourselves, is still longer than *Homo sapiens* has yet existed on this planet.

To me, the prospect that the entire Earth may have even mildly resembled Jupiter's frozen moon Europa, and that this may have happened not at the beginning of life on Earth but smack dab in the middle of life's evolutionary history here, is truly mind-boggling. Moreover, the fact that this idea and the evidence supporting it have been brought together only in the last decade or so suggests that our planet may yet hold many fascinating secrets just waiting to be "unearthed."

Certainly, following the great freeze and thaws, life found itself in constant turmoil. With lines of communication (that is, water) literally frozen around the world, the life that survived could have begun to diverge from its distant cousins elsewhere. Where once symbiosis ruled the planet, perhaps competition for scarce resources was now the order of the day. Metaphorically, at least, we may have come to the end of the garden of Eden, where the Earth's bounty existed to be shared by all, to a time when "Kill or be killed" was the rule.

But all of this is immaterial to our oxygen atom, which following the last great thaw would have once again resumed its familiar life cycle, moving from atmosphere to organic life to water to rock to carbon dioxide and back again. By 600 million years ago, the oxygen levels on Earth were certainly close to their present value, and the Earth's biosphere had become essentially identical to the Earth that humans would later grow so fond of. Plants and animals, including some that are definite direct ancestors of animals that walk (or swim) the face of the Earth today, began to abound on continents that were moving apart to reach their present configuration, which is of course still transitory.

But it appears that without some more serious shuffling, the planet would not really have been ready to reach its present plateau, where life is now apparently increasing the carbon dioxide content of the atmosphere, as opposed to fixing carbon in the ground and on the Earth's surface. It seems pointless to argue whether it is a good or bad thing for the Earth that intelligent life eventually arose and is now overseeing one of the greatest extinctions of all time. Nature is neither good nor bad. Nor does nature care about individual life, or even whole civilizations. Life is simply a matter of being in the right place at the right time, and likewise death tends to be a matter of being in the wrong place at the wrong time. One may devote oneself to wondering whether this is a result of chance or predestination, but I don't see much value in the whole debate.

Indeed, while we conceivably might someday render our home unsuitable for our continued existence, it has been uninhabitable before, and it certainly will be again, whether or not we hasten the process. The immediate future may matter to us, but the Earth has already seen more devastation than we are likely to inflict. All we can do, it seems to me, is make the best of what we know will ultimately be a bad situation. In this sense, as always, fortune favors the prepared mind.

In discussing the past 3 billion years, we have paid scant attention to the great messengers from heaven that ruled the early Earth, bringing both life and death. This is because the frequency of large comet and meteorite bombardment fell off exponentially with time, thankfully. As I have described, an object larger than about 300 kilometers would vaporize the oceans completely, and heat the crust to more than 1,000 degrees Celsius. Such sterilizing bombardments occurred with regularity early in the Earth's history, but there is no evidence that any such event has occurred in the last 3.5 billion years. Life appears to have survived continuously throughout this period, at least in part because the odds were in its favor.

But the predicted rate of large-scale bombardments today is not zero. Every 100 million to 300 million years or so, on average, we are guaranteed that a large object will slam into this planet, creating havoc with almost every living thing. I have also mentioned that coming from the asteroid belt there are estimated to be 1,000 to 2,000 objects larger than 1 kilometer in size that are on such possible Earth-crossing orbits. Statistical arguments suggest that roughly every 300,000 years a 1-kilometer object should collide with the Earth, and a 10-kilometer object should collide on average every 30 million years or so. An object slightly bigger than this, between 10 and 20 kilometers across, is thought to have been responsible for the famous extinction of the dinosaurs some 65 million years ago.

Of course, not all large-scale extinctions can be blamed on comets or asteroids. It might be a coincidence that perhaps a dozen mass extinctions have occurred over the 540 million years since the dawn of the Cambrian era and this is also the number of large objects one might predict would have collided with the Earth during such an interval. After all, extinction is an essential part of life, as far as we know it. The sulfur bacteria are still here, and we may be accustomed to thinking that we are invincible as a species, but in fact the vast majority of species that have ever walked the Earth or swam in its seas are now extinct.

Later in this book I will address our own mortality as a species, but it is not too early to begin to get used it. Nature casts aside species as it casts aside individuals. Some are replaced by evolutionarily more adept versions, some die out because of long-term changes in the Earth's biosphere, and some perfectly functional species get killed by accident. But nature has a kind of cosmic insurance policy. The beneficiaries of the death of a species are other species that may have lived in its shadow, and that can grow to fill the unoccupied niche. Mammals are just such a species. The demise of the dinosaurs opened the way for mammals to grow in size and diversity, and the rest is history.

The largest extinction known on Earth occurred about 250 million years ago, in what is called the Permian era, when the supercontinent of Pangaea had begun to form. No one knows the precise cause, or causes, of this catastrophe, but within a short period perhaps 96 percent of all species then alive on Earth became extinct. Nothing in the recorded history of life before or since has ever involved such a pervasive demise. Most certainly a cooling climate played at least a partial role. Remarkably, however, no evidence exists that this wholesale change in the cast of characters living on this planet was induced by any single traumatic event, underscoring once again that there are many ways to die.

The great Permian extinction may have simply been a case of bad luck, with many little factors adding up at the same time, just as the great freeze that may have initiated Snowball Earth was induced by a particular combination of a cool sun and a single supercontinent located around the equator. If the Permian extinction was indeed merely bad luck, these massive deaths would have been but a footnote to the history of our oxygen atom. Most certainly the specific pathways by which oxygen was absorbed, transported, and eventually excreted may have changed, but these are peripheral. The ultimate input and output of oxygen as part of life's cycle have remained more or less the same for over 2 billion years. The reason is not a lack of imagination on nature's part. It is rather that one apparently cannot do any better with the materials available.

Two hundred and fifty million years later, give or take a few million, an evolutionary demise on a much smaller scale would take place. It would hardly be worth mentioning in this broad-brush history, were it not associated with an animal that many people take for granted as the pinnacle of the evolutionary tree.

No less than 20 different species of hominids have been uncovered over the years, going by exotic names like *Ardipithecus ramidus*, *Australopithecus africanus*, *Paranthropus robustus* (although obviously not all that robust), and eventually *Homo ergaster*, *Homo erectus*, and *Homo neanderthalensis*. Many of these shared the planet for periods perhaps 50 times longer than the time span since the most recent species, *Homo sapiens*, arose out of Africa. Why, over the course of 5 million years, some died out and others did not is not known, except perhaps for the last extinction, in Europe, of *Homo neanderthalensis*. Although most of the previous hominids had shared the landscape in relative peace, it is likely that *Homo neanderthalensis*, a large-brained, utilitarian, but apparently peace-loving brute, succumbed to a predator without sharp fangs or claws. His demise was probably hastened by a close relative, one who may have shared with Neanderthal man a gift that many scientists and religions had assumed was given to only one species on Earth: self-awareness. Neanderthals buried their own, perhaps originally to avoid scavengers. But sometimes these burials were performed with obvious tenderness. In one neatly arranged grave, a bouquet was left with the body. While they may not have had explicit language, they certainly knew how to say it with flowers.

The awareness that another "soul" had passed away probably requires an awareness of self. But as profoundly important as this primitive sentimentality was, *Homo neanderthalensis* appeared to go no further. *Homo sapiens* is, as far as we know, the first creature in almost 4 billion years of life on this planet that developed an explicit spirituality. Burial goods, animalistic art, exquisitely crafted tools — all these were the unique products of the mind of our species, developed along with the notions of God, magic, and evil, sometime in the past 40,000 years. And our spirituality has been the root of most of the organized violence in recorded human history. *Homo sapiens* has a capacity for destruction unparalleled in the history of life on this planet. Wherever ancient human settlements spread, other species were in danger. *Homo neanderthalensis* was probably just one in a long line of victims.

The thought that *Homo sapiens*'s self-awareness may not have been unique is perhaps shocking. When the species is seen on a continuum of all the hominid species that sprang up around the world over a 3-million-year period, rather than as any kind of giant step forward, it seems more a simple twist of fate that *Homo sapiens* ended up the sole survivors. What if it had been the Neanderthals? Evidence from the one place in the world where it is clear that the Neanderthals and *Homo sapiens* co-existed for some time, the Levant, demonstrates that the two ended up producing very similar tools. Who is to say that Neanderthals might not, at some point, have developed a society, an art, and a culture? But in this case, what does this notion do for the idea of a human made in the image of God? And if such a God had preordained for humans to arise, having them do so at the expense of Neanderthals seems a haphazard way to go about it. *Homo sapiens* appears as one of many branches on a hominid tree, its main distinction simply being that it is the only branch to still be sprouting buds.

Indeed, the significance of these long-dead cousins is that they put us in our place. We are far from being the capstone of a mansion built by evolution. It is as if many parallel hallways existed, some of which ultimately led onward, and others of which went nowhere at all. The hallways are long, and it is impossible when entering them to know where they lead. Like all other species, we are an evolutionary experiment in a process that involves apparently random events rather than a logical progression. At 40,000 years, we have been around for less time than any other species from antiquity that is encoded in the fossil record. It is premature to jump to conclusions.

Nevertheless, when the first *Homo sapiens* looked up at the stars and wondered what she was looking at, the future of the biosphere in which our oxygen atom resides changed forever. As a species, we are the first to have the capacity to alter our environment, both locally and globally. Our burning of fossil fuels, for example, appears to be affecting the level of greenhouse gases with as yet undetermined consequences. Ultimately the future of this planet may lie in our hands, as may the future of our own species. If we are lucky, we may be called upon to use our incredible skills and creativity to rescue the Earth. Or it may be that the future is completely beyond our control. Chance events may determine our future as surely as such events set the stage for our present existence.

Indeed, this is a good opportunity to return to that singular event 65 million years ago that happened to change the course of evolution in our favor. Its impact on our existence is, however, not the only reason for such an excursion. Rather, unlike the subsequent course of human evolution, the event that marked the demise of the dinosaurs also had an effect on the hero of our story, our oxygen atom. In comparison with the Permian extinction and many of the others that followed it, the event that killed the dinosaurs and made way for mammals was truly cataclysmic. Moreover, abundant evidence now indicates that this extinction had an extraterrestrial origin.

Examination by geologists of rock throughout the world has long suggested that 65 million years ago a dramatic change took place in the Earth's biosphere. At an obvious point in the layers of sedimentary rock, a sudden change in the fossil composition indicates a massive extinction. In this case, however, there is even more dramatic physical evidence of a global change. The boundary between the layers created before this time (a period called the Cretaceous) and the layers above it (created in a period called the Tertiary), is known as the *K/T boundary* (K stands for the German spelling of Cretaceous). Associated with this boundary is a layer of clay, which varies in size from place to place, upward of 1 centimeter in thickness. In the late 1970s several research groups decided to examine this clay layer and found a completely unexpected result. Recall that in the Earth's core are elements called *siderophiles*, whose presence in the crust is small because these elements would have tended to follow iron and nickel into the core of the planet during its early molten stage of formation. In meteorites, which never underwent this kind of segregation, these elements are much more abundant than in the Earth's crust. As the siderophile abundance in the K/T clay layer was examined, a striking anomaly was discovered. The element iridium was found to have an abundance 10 times greater in this clay region than in the rocks above or below. Moreover, in the clay layer grains of a special type of quartz were found that could only form under high temperature and pressure.

Taken together, these observations provide highly suggestive evidence that a large amount of extraterrestrial material was transported to Earth at this time, in a way that would produce high temperatures and pressures, and in a way that might affect the Earth dramatically enough to kill off many of the life-forms then extant, including the dinosaurs. The obvious possibility is a large asteroid impact. It would have had to be large enough to have produced a worldwide catastrophe, but not so large as to sterilize the planet completely. A 10-kilometer-size object would fit the bill.

Such an object would produce a crater of about 100 kilometers across, which you might expect would be something that is hard to miss, even after 65 million years. Of course, since much of the Earth is underwater, if the impact had occurred in the ocean crust it could easily have remained invisible, or even have been obliterated by being subducted under one of the continental plates in the 65 million years since that time. Moreover, erosion from wind and water on land can also dramatically change the look of things. Nevertheless, scientists searching for a crater to validate this hypothesis were lucky, or perhaps half lucky. Lying half in the ocean and half on land, a huge crater 200 kilometers across was ultimately discovered off the coast of Mexico's Yucatán peninsula. Was this the first direct proof of the great dinosaur killer? While the evidence is, after all, purely circumstantial, it has continued to grow with time. And in this case, the glove definitely fits.

What would have happened that fateful day (or night) will never be known exactly, but we can surmise, referring to general laws of physics, what might have happened. This huge rock may have been dislodged from its position in the asteroid belt a thousand years earlier. Each subsequent orbit around the sun would have brought it closer and closer to its ultimate target. Eventually, it would have found itself hurtling silently through space toward the Earth at a speed of perhaps 25 kilometers per second. At such a speed the distance between the Earth and moon can be traversed in just over 4 hours. No forewarning of impending doom would be given. Even with our current telescopes trained on the heavens, we have detected only about half the 10-kilometer-size objects currently thought to be on possible Earth-crossing orbits. The dinosaurs had no such remote sensing apparatus. Grazing on what was surely, somewhere, a cloudless day, they would have had no inkling of what was about to happen.

As the asteroid first entered the Earth's atmosphere, tremendous heat and pressure would have begun to be generated by the drag force of the air. The rock would have glowed red hot, perhaps shattering, and begun to vaporize, so that not all of the asteroid would have made it to Earth. But most of it would, and when it struck, an explosion unlike any witnessed on Earth since that time would have occurred. Imagine a crater 200 kilometers across — larger by almost a factor of 2 than the width of any of the Great Lakes — being formed in just seconds. The heat generated at the impact site would have melted or vaporized much of the surrounding rock.

In such a collision, the impact energy would have been great enough (creating an earthquake of magnitude 12.4) to knock material from the crater right out of the atmosphere, causing it to fall back down to Earth on the other side of the planet! A simple estimate of how much material is removed when a crater 200 kilometers across and 1 kilometer deep is excavated yields roughly 50,000 billion tons of dirt and rock. This is enough material to bury all of the landmasses on Earth under almost half a foot of material.

But mere burial was not all that was in store for the planet. When rocks ejected from the crater are shot into space and return, they too heat up in the atmosphere. Everywhere on Earth a red-hot rain of material would have fallen, with a brightness 10 times that of the sun, setting off global fires and scorching much of the surface of the planet. Furthermore, even if burial on land was avoided, burial at sea might not be. If such an impact occurred, even partially underwater, the tidal wave sent around the planet would have devastated many coastal areas for thousands of kilometers. Rubble found in Haiti, for example, is thought to come from the tidal wave initiated by the impact that caused the Yucatán crater.

Much of the dust would not have fallen right back to Earth, however. It might have remained in the atmosphere for a year or two. During this time, it would have obscured the sun, turning day into night. The combination of freezing temperatures that might result and the lack of direct sunlight would kill off many of the photosynthesizing plants, destroying the bottom part of the food chain.

Finally, if all of this were not bad enough, the huge amount of water thrown up into the atmosphere would have led to torrential downpours in the weeks and months following the impact. This would have been more than merely a hard rain. The heat imparted to the atmosphere would have allowed many new chemical reactions to occur, including reactions with nitrogen, producing nitrous and nitric acid, among other compounds. The rain that fell would thus have been acidic, harmful to many forms of plant and sea life.

Finally, in the long term, the shock of this impact on the Earth's crust could have provoked massive volcanoes to erupt throughout the planet, affecting nearby locations and pumping huge amounts of carbon dioxide and other gases into the atmosphere. Among other things, this would have created a short-term greenhouse effect that could have resulted in scorching temperatures in the years following the deep freeze induced by darkness. The poor dinosaurs wouldn't have known what hit them.

Our oxygen atom could have played many different roles during this catastrophe. It could have helped combust trees and other organic materials, feeding the flames of global forest fires and converting to carbon dioxide in the process. It could have combined with nitrogen in the atmosphere to become a part of the worldwide acid rain. If it were in the ocean at the time, it could have been part of a tidal wave that might have obliterated a tropical island previously lush with life.

Following the event, with reduced photosynthesis and presumably much larger carbon dioxide fractions in the atmosphere, our oxygen atom might have lingered, unused, in the atmosphere for some time, or have dissolved in the ocean water, perhaps to power some respiring sea life that survived the global trauma. Or, as part of the acid rain, it might have leached material from the rock below, oxidizing the materials, and have been carried with them through streams to the ocean, where it might have settled, been subducted tens of millions of years later, and reappeared once again in a volcanic burst. The new world it would experience would have been noticeably different, in some senses, than the world before the impact. The climate would have reverted to a more hospitable one, but the animals taking advantage of the sun, water, and oxygen would have been dramatically different ones, for the most part. Mammals would now have begun to rule the planet, filling the niche left by the dinosaurs, and paving the way for the humans to come.

The K/T impact was probably the last time in the history of the planet that our oxygen atom's life cycle was altered significantly. At some point after this event, oxygen would again revert to its traditional role in the biosphere, participating in all the processes of life, from the excretions of the smallest insect to the deep breaths of the largest whales, day in, day out, over the eons.

For each oxygen atom on Earth, the last 65 million years may have been relatively unremarkable. Subsequently we shall see that, even from the perspective of an oxygen atom, the future appears destined to be exciting again. Indeed, unless something emerges to alter the probable future, our atom's experience during the K/T transition will have been tame by comparison.

Nevertheless, the life cycles of oxygen over the past several thousand years have been of particular interest to a biped species with the gift of language, literature, and mathematics. Throughout this story, I have tried to present the picture, whenever possible, from a perspective centered on the oxygen atom itself. However, because this precise point in our narrative relates so directly to our present existence, I am going to switch perspectives in the next chapter. When seen through the eyes of an active participant in the present human experiment, the recent history of oxygen atoms on Earth takes on a completely new significance and flavor, and offers a few surprises.

16.
THE BEST OF TIMES, THE WORST OF TIMES

I am an estuary into the sea.
I am a wave of the ocean.
I am the sound of the sea.
I am a powerful ox.
I am a hawk on a cliff.
I am a dewdrop in the sun.
I am a plant of beauty.
I am a boar for valor.
I am a salmon in a pool.
I am a lake in a plain.
I am the strength of art.

AMHAIRGHIN, DRUID POET,
SOMETIME BEFORE 400 A.D.

I t may seem like a non sequitur, but whenever I think of the lives of our atom over the span of human history, indeed when I think of human history itself, the city of Venice always comes to mind. Anyone who has ever visited can never forget the experience. For me, it is a city of romance and intrigue, yet it also harbored the first modern scientist, Galileo. The sounds you hear at night can be either the rocking of boats in the water or the sound of fine Italian shoes clacking against the stones in the alleyways. And anyone who has ever walked these alleys will inevitably ask, at one time or another, usually moments before realizing they are lost, "Haven't I passed this way before?"

Venice is a city built in every sense around water. The most direct way from any place to any other place is via the canals, not by foot. Radiating from its many piazzas are innumerable narrow passageways, none of which ever proceed in a straight line for very long. Because these provide the only visible breaks between the buildings, it is impossible to see directly what might be a stone's throw away. Moreover, while each alley has its own charm, they all look superficially alike, so that it is frustratingly difficult for a new visitor to know exactly where he is at any time. You can be within a hundred steps of your destination and not know it. At the same time, there is never great cause for concern, because eventually, usually before your legs wear out, almost any route will ultimately meander past the place where you were originally heading.

Venice could serve as a metaphor for the cycles of life on Earth. The history of each individual or each atom on Earth is unique, but at the same time the story of any one is fundamentally the story of all. What may appear to be quixotic twists and turns of fate yield an ultimate progression unforeseen at the beginning, or even during the voyage. To change metaphors slightly, it is as if we are all threads in a grand Venetian tapestry. Each thread simply moves behind or in front of its compatriots, in a manner that must appear arbitrary when viewed under a magnifying glass. Each progression differs in almost every detail from that of its nearest neighbor, but in bulk they are all more or less equivalent. Yet when viewed together, the threads weave a remarkably rich and intricate picture.

This book arose from the recognition that we are all star-children. Every atom in our body was once inside a star that lived and died so that one day we might be born. But at the same time, we can lose sight of the fact that we are equally Earth-children. Each atom in our body is only a temporary visitor, for minutes to years, depending on its particular location. Thus far in this history I have focused on the lives of an individual oxygen atom beginning at the dawn of time. Its story may not correspond precisely to that of any particular atom actually on Earth today, but may instead, like many a literary hero, be considered a composite drawn from the history of many different individuals. On the other hand, the number of atoms on Earth is immense, so that any history one might imagine is likely to have actually occurred for at least one of them.

The great Italian writer and chemist Primo Levi closed his semiauto-biographical book *The Periodic Table* with a delightful short chapter that traces the history of a carbon atom from the moment it was dug up in a limestone rock in 1840 until the time of his writing, in 1975. I just reread this piece because I wanted to remember how he envisioned this brief 235-year interval, and because I hoped that by mentioning it I could head off the mail from readers who might wish to remind me of this classic work. I was surprised and delighted to find that Levi made almost precisely the same claim that I just made, although his is stronger. He argues that given the number of atoms on Earth (in this case carbon atoms), it is guaranteed that any history one might invent, no matter how capricious, will have already occurred. As you can see, I am less brave. As a cosmologist, I am used to hedging my bets with the phrase "is likely to."

In any case, Levi was probably so excited about providing a delightful end to his carbon atom's tale, which I will not divulge here, that I fear he missed a golden opportunity to take his argument to its logical conclusion. That is what I want to do here, in the case of our oxygen atom.

Namely, I want to depart from our linear history and focus not on the history of one specific atom, but rather on the near infinity of histories encompassed by, say, all the oxygen atoms in the breath of air you are taking as you read this. This is a unique time in our narrative. We are verging on the present. These different atomic histories take on a special significance when they reach out and touch us directly. Thus the right time for such a discussion is now. Soon it will be too late to return.

It is a unique feature of the statistics of very large numbers that allows both Levi and me to claim that our fabrications might somehow reflect reality. These very statistics also yield remarkable new insights about our connections to the past.

Consider the following: How many oxygen atoms are in each breath you take? This is simple to answer. Say each breath of an average human being comprises about half a liter of gas. A liter is 1,000 cubic centimeters, and at room temperature and pressure the density of air is such that 1 liter of air encompasses about 1.5 grams of material. Now, for any gas at room temperature and pressure, 1 liter corresponds to roughly 1/20 of a *mole* of gas molecules. A mole of any substance contains precisely the same number of molecules, about 6×10^{23} (a 6 with 23 zeros after it) molecules. Thus there is about 1/20 of this amount, or about 3×10^{22} molecules of gas in each liter of air. Since molecules of oxygen and nitrogen contain 2 atoms apiece, there are about 6×10^{22} atoms in a liter of air. Since oxygen is roughly 1/5 of all the atoms, this makes about 1.2×10^{22} oxygen atoms in a liter of air. Thus in each breath of air of about half a liter, there are about 6×10^{21} oxygen atoms. This is a lot of atoms. So many, in fact, that they span many, many different histories so that we can argue that no matter how unlikely any scenario is for any given atom in a breath full of atoms, *some* atom in each breath may have experienced it.

We can put some mathematical teeth in this assertion by following our line of argument a little further. First, we must estimate how often it takes for an average oxygen atom to be recycled through some living system in the Earth. There are lots of independent ways I have tried to estimate this, and thankfully they all give about the same number. Here is one example:

An average forest cycles about 2.6 kilograms of new organic material in each square meter of forest each year. About 80 percent of this material is used to respire carbon dioxide and water back into the atmosphere, and 20 percent is stored. Assume this value for new organic material production is a reasonable order-of-magnitude estimate for the amount of organic material produced per square meter per year over the non-desert, non-icecap parts of the world (including the ocean, where photoplankton are active photosynthesizers). Next, using the fact that there are about 400 million square kilometers of such land and water on Earth, we can estimate that roughly 1,000 billion tons of organic material is produced each year by life.

How much oxygen is there in the atmosphere for use in such production? We can cite the simple fact that the pressure of air on the surface is about 15 pounds per square inch. Thus the total mass of air above each square inch on the surface is about 15 pounds. Turning this into metric units, one gets a mass of about 1 kilogram per square centimeter. Taking the total surface area of Earth to be 5 billion billion square centimeters, one gets a total mass of 5 billion billion kilograms. Since about 1 in 5 molecules in the atmosphere is oxygen, this gives about 1 billion billion kilograms (or about 1 million billion tons) of oxygen in the atmosphere.

Finally, let's review the process of creating organic material in photosynthesis. In photosynthesis, each atom of oxygen produced when water is split yields 2 protons, whose energy is used to fuel the generation of 1 molecule of ATP. It takes 18 ATP molecules to generate 1 molecule of glucose, whose weight is equivalent to about 12 oxygen atoms. Thus in creating 1 gram of organic material, roughly 1.5 grams of oxygen is produced. Based on the above estimates, in about 650 years organic processes could generate every oxygen atom in the atmosphere.

Of course this estimate is fraught with uncertainty and possible error. Some of the oxygen produced by life will not go to the atmosphere, but will remain with organic sediments and dissolve in water in the oceans. There is a lot of oxygen stored in these systems, compared with the oxygen in the atmosphere, just as there is roughly 100,000 times as much carbon stored in rock and organic sediments as there is contained in carbon dioxide in the atmosphere. Nevertheless, we have independent reasons to believe that a time period of centuries is reasonable for the complete recycling of an average oxygen atom in the air.

For example, I learned from Primo Levi that every atom of carbon not tied up in rocks is recycled through life in a time period of about 200 years. Now, there is much more oxygen than carbon dioxide in the atmosphere, so one might figure it would take longer for all the oxygen to recycle. But when one takes into account the amount of carbon in accessible organic materials as well as in the atmosphere, this makes the total amount of usable carbon in the biosphere not that different from the total amount of oxygen in the atmosphere.

Finally, living materials on the surface of the Earth are essentially fully oxidized on a time frame of years to decades, on average. A geologist colleague of mine claims he makes this point to his classes by suggesting to them that if this weren't the case, it wouldn't take long before we were buried by our own grass clippings.

So let us, for the sake of argument, now imagine that the oxygen atoms we are now breathing are continually redistributed again throughout the atmosphere on a time frame of centuries. This means the molecules in every breath we inhale will, over the course of the next millennia, if not the next century, become redistributed uniformly throughout the atmosphere again. If this is the case, then we are more connected to our past than any of us might care to imagine.

To use a well-known example, let us consider the moment Julius Caesar was killed, and exclaimed in his dying breath, "Et tu, Brute?" The estimate above allows us to demonstrate, by one of several different routes, that every time you breathe in, there is a reasonable likelihood at least one of the oxygen atoms you inhale was contained in Caesar's last breath.

First, we know how many molecules were exclaimed by Caesar, because we worked this out earlier. An average breath contains about 6×10^{21} atoms. Since Caesar undoubtedly heaved a big sigh with his last breath, let us assume it contained 4 times as much, about 2×10^{22} atoms. We also can work out how many oxygen atoms there are in the entire atmosphere, using our estimate above of 1 million billion tons of oxygen gas. This works out to contain about 4×10^{43} atoms of oxygen. This means that the fraction of the total atmosphere today made up of oxygen atoms in Caesar's last breath is 5 parts in about 10^{22}, a *very* small fraction. But if in each breath you take (even without a sigh) you breathe in 6×10^{21} oxygen atoms, the above fraction implies that you are taking in on average about 3 of the oxygen atoms in Caesar's last breath in each of your breaths!

The mathematician John Allen Paulos reached a similar conclusion using an alternative, probabilistic argument: If the probability that a molecule you breathe in came from Julius Caesar is about 2×10^{-22} (his assumptions were slightly less optimistic), then the probability that the first molecule you breathe in was *not* from him is about $(1 - 2 \times 10^{-22})$, or very close to unity. The same is true for the next molecule, and so on. Thus the probability that *all* the molecules in your breath are not from Caesar is the *product* $(1 - 2 \times 10^{-22}) \times (1 - 2 \times 10^{-22}) \times (1 - 2 \times 10^{-22}) \dots$, with the number of terms equal to the number of molecules in your breath, which he assumed to be about 2×10^{22}. This product turns out to be less than 0.01, implying that the probability that none of the molecules in your breath came from Caesar's last breath is less than 1 in 100.

This is the good news. You are truly a part of noble history! But by the same token, you are a part of ignoble history. If there is likely to be a molecule from Caesar's last breath in every breath you take, there is also likely to be a molecule from *every* breath that Caesar took in every breath you take. The same goes for Cleopatra. And if the recycling time for air in the atmosphere were shorter than a century, the same may go for Adolf Hitler. In this sense, there is likely to be a molecule coming *from every breath* from *every person* who ever lived (up until too recent a time for these molecules to have been recycled through the atmosphere to you) in every breath you take!

But it gets more interesting. If oxygen, during respiration and photosynthesis, is alternately partnered up with hydrogen as part of water, then there is a reasonable likelihood that at some time every oxygen atom in your breath was a part of a water molecule. But this water molecule has some nonzero probability of having been contained in the excretion of someone who lived before on Earth. You thus may be breathing remnants of the urine, and semen, of many of the people who have come before. The sweat of your parents' intimate couplings, indeed perhaps that associated with the moment of your conception, may be contained in the glass of water you drink today. In fact, we need not confine this argument to people: horse urine, pig feces, the whole shebang! We can carry this argument back to the beginning of life on Earth. One can calculate, by the same assumptions, there is a reasonable chance that sometime while you are alive, you are breathing in some atom excreted by at least one of every species that has ever been alive on Earth since the time oxygen first built up in the atmosphere!

If this is too much to bear, let me remind you that the estimates I have made here are very rough, so it could be that you are not guaranteed to have a molecule from Caesar's dying breath in the air you breathe. Rather, it could simply be possible, or maybe even merely barely plausible. But the bottom line is the same in any case. Each oxygen atom in the breath you are taking at this moment has had a unique history. Some histories are exotic, and some are not. I believe that at least one corresponds to the history I have written thus far. But the set of histories embodied in all the atoms in a single breath is so large that their stories could not be told even if every page of every book ever written in human history were devoted purely to telling them.

The ancient poetry of Amhairghin I quote at the beginning of this chapter underscores how you are connected, each time you breathe in and out, to almost all the rest of life on Earth, today, and in the past. And before the Earth was formed, to the stars . . .

And by the same token, you are equally connected to the future. Sir Arthur Stanley Eddington, the renowned British astrophysicist whose admonition to those who disagreed with him about the processes taking place inside the sun I quote at the beginning of chapter 8, described in 1935 a thought experiment similar in spirit to the examples I have presented here: "Take a cupful of liquid, label all the atoms in it so that you will recognize them again, and cast it into the sea; and let the atoms be diffused throughout all the oceans of the earth. Then draw out a cupful of sea-water anywhere; it will be found to contain some dozens of the labeled atoms."

Does this example point to our own indelible, if anonymous, imprint on the future? Eddington surely thought so, for he argued that because of it, "We can read a literal meaning into Macbeth's words":

> Will all great Neptune's ocean wash this blood
> Clean from my hand? No, this my hand will rather
> The multitudinous seas incarnadine . . .

17.

THROUGH
A GLASS
DARKLY

*"Men's courses will foreshadow certain
ends, to which, if persevered in, they
must lead," said Scrooge. "But if the
courses be departed from, the ends
will change. Say it is thus with what
you show me."*

CHARLES DICKENS

Off the coast of the Yucatán peninsula, half submerged,
half worn away by tens of millions of years of erosion
and weathering, and ultimately buried under a tropical jungle, lies a har-
binger of the future. Going by a name that is pronounced like "chicks
you love" said with a plugged nose, Chicxulub is the largest crater
known on Earth. Originally estimated after its discovery in 1991 to be
about 200 kilometers across, it may be, some topographical evidence
suggests, as much as 50 percent larger.

No matter. Either way, the object that created the crater 65 million
years ago was deadly. It helped kill off the biggest land animals ever to
walk the face of the planet, along with a host of other species on land and
in the sea. I say "helped" because there is evidence that this interplanetary marauder may have in certain cases merely completed a job that nature was already accomplishing on its own. Of course, we will probably
never know what would have transpired had this asteroid not intervened.

It is, as I have explained, an unavoidable fact of life that in an evolving biosphere most species of life are destined for extinction.

In any case, as sure as the sun is shining, somewhere out there lies another, perhaps bigger object that is being inexorably driven by the laws of classical mechanics and the force of gravity on a trajectory that will one day cause it to smash headlong into our blue-green planet. That much is certain. Whether our species will live long enough to experience this particular trauma, or the many others that will inevitably follow, is an open question.

We are now about to embark on a tale of the future of our oxygen atom. I am painfully aware in this regard that as much as the past history of our atom has involved both speculation and inference and thus is subject to some ambiguities, a discussion of its future involves speculations at a whole different level. At the end of the nineteenth century the triumph of a mechanistic worldview led some to believe that the future could be predicted as surely as one could follow the movements of the gears and wheels in a gigantic clock. A century later, our worldview has matured. Aside from the inevitable uncertainties associated with the quantum mechanical behavior of atomic systems, we now understand that even in purely deterministic classical evolution, small variations in initial conditions can result in dramatic variations in the ultimate outcomes. The notion, for example, that a butterfly fluttering its wings in the Midwest may ultimately cause a hurricane in the Northeast has by now permeated the public's consciousness, however unrealistic such a possibility might actually be. Nevertheless, we now recognize that classical systems can be truly chaotic, wherein infinitesimally small variations in initial trajectories can be amplified tremendously to result in dramatically different later ones.

In fact, we now understand that even that paragon of predictability, the motion of the planets in our solar system, is chaotic. Over the course of millions of years, small perturbations — so small that one could never hope to account completely for their presence at any one time — will build up to result in measurable alterations of the motions of the planets, including the Earth.

In such a situation one might be tempted to give up hope of predictability. But this alternative is equally fallacious. Physicists every day confront situations where the exact outcome of a single experiment can never be predicted with certainty. We can know, however, with absolute certainty the statistical likelihood of all the different possible outcomes. Similarly, even if we were to locate and catalog all of the 1,000 or so 1-kilometer-size objects in the solar system currently on near-Earth-crossing orbits — something that could probably be achieved with several decades of dedicated observing — it is still not likely that we would be able to absolutely predict when a collision with Earth will occur. Nevertheless, we can predict with very high accuracy that one such collision should occur, on average, within about 300,000 years.

Such a collision is not likely to cause a global extinction, although it will certainly be traumatic. The object that created Chicxulub is thought to have been between 10 and 20 kilometers in diameter. We expect there are about 10 to 20 such objects presently in the solar system on near-Earth-crossing orbits, and that Earth's average interval between collisions with objects this large is about 30 million to 100 million years. While this is soothingly far in the future, when the next such collision happens — and it will — the consequences for life on Earth will be devastating.

Or will they? As hard as it may be to predict the future based on the random motions of uncaring objects, the introduction of intelligence into the equation means all bets are essentially off. The dedicated international planning combined with luck and macho determination that allowed the Earth to survive two separate asteroid impact events in movie theaters recently was undoubtedly far-fetched in the extreme. But it is not completely implausible that with perhaps a decade's notice, an intelligent civilization might devise some mechanism of staving off the catastrophe.

The same might be said for the likelihood of survival through any of the other calamities I am about to describe. Scrooge addressed the Ghost of Christmas Future with the famous admonition: "Before I draw nearer to that stone to which you point, answer me one question. Are these the shadows of the things that Will be, or are they shadows of things that May be, only?" Encouraged by the latter possibility, Scrooge became a changed man and, we are told, thus changed the future. Whether we will be wise enough to benefit from advance warning of future crises, from the short-term possibility of global warming to the long-term evolution of climate on Earth as the sun evolves, is anyone's guess.

In one sense, none of this matters for our oxygen atom. We may labor under the fallacy that somehow this is a special time in the Earth's history, and that we, as an intelligent species, are the pinnacles of evolution and the ultimate guardians of the fate of our planet. But it is unlikely that in the long term, matters will be under our control, or that we will even be here to control them. Our oxygen atom has survived devastation that wiped more than 95 percent of living species off the face of the planet, just as it has survived the brutal formation of the Earth itself, and the death of the star that ultimately carried it to our forming solar system. Whether or not any life survives on Earth is unlikely to be of major consequence to our atom's future.

At the same time, the word *unlikely* is vital here. It is not guaranteed that our descendants, whatever form they may take, will not affect the ultimate future of our atom. When I began this book, for example, I was convinced that the lives of our atom would extend well beyond the existence of consciousness. In the intervening period my views have changed somewhat. Now I am only *almost* certain.

I am by nature skeptical. As such, I cannot say that I am personally optimistic regarding the future of the human species, although I am optimistic that certain aspects of our consciousness may survive beyond the demise of the Earth. Nevertheless, I will forgo any such pessimism here. As I describe the future of our oxygen atom, I will attempt to follow the guide of nature, and alternate randomly between optimistic and pessimistic developments. Whenever necessary, I will attempt, out of charity, to give the benefit of the doubt to humanity.

The somewhat eclectic astrophysicist J. Richard Gott III, at Princeton, has obtained some notoriety for espousing a principle that is without any distinct physical basis, and fundamentally grounded in ignorance, although mathematically sound. His Principle of Copernican Time is based on the notion, which is periodically false but often true, that there is absolutely nothing special about any particular set of circumstances. Thus, for example, we know that we are not located at the center of the universe, or even the center of the solar system, as Copernicus argued. Moreover, we now know that we are not located in the center of our galaxy, but rather in a remote suburb, and that our galaxy is located in a group of randomly distributed galaxies, amid a much bigger cluster of galaxies, and so on.

Gott has pointed out that one can also apply this theory to the notion of time by making the claim that, *without any a priori knowledge of the details*, we should expect that the lifetime of any system is not likely to be orders of magnitude larger than the amount of time it has already existed. This claim is simply based on the notion that if some system ultimately survives n years, one is statistically unlikely to stumble upon it either extremely close to birth or death. Thus the fact that I am 46 years old suggests, even if you don't know the average life span of a human being, that I am unlikely to live, say, 1,000 years more. In fact, assuming things are about as random as they can be, one can put quantitative estimates, given the fact that I have thus far attained this age, on how likely it is that I will live another 46, 460, or 4,600 years.

One can apply this kind of reasoning to all sorts of different systems, and Gott claims great success, for example, in predicting the lifetime of Broadway shows based on their current run. One should, of course, not read too much into this. The application of statistics to the human condition is, as we all know from our experience with electoral predictions and outcomes, subtle. Moreover, Gott's result depends crucially on knowing absolutely nothing about the system in advance. If one does, then Gott's argument is inappropriate. Nevertheless, a straightforward application of his principle suggests that if *Homo sapiens* has roamed the planet for only 40,000 years, then it would be very surprising if we were to survive as a species for more than about another million years or so.

I certainly believe that this estimate is plausible, but at the same time, it contains virtually no useful information. Will we die out dramatically, leaving no descendants in any form? Or will we take advantage of certain opportunities as they arise to evolve into a life-form that is distinctly different from our current one? On good days, I believe the latter. For example, two developments of the past 25 years, the invention of digital computers and the ability to sequence DNA, suggest to me one interesting possible course for humanity.

I see no obstructions to the creation of intelligent, self-aware, self-programmable, computing machines. If this occurs, these machines will have a tremendous evolutionary advantage over purely biological machinery, which is constrained to improve our human computing capabilities much more slowly. At the same time, our ability to sequence the entire genome of human DNA means that ultimately what we consider life, and biology, will change. We will, I believe, soon be able to manipulate living systems on scales currently unthinkable.

It seems to me that this combination of technologies has one logical outcome. Humans, if they are to compete with the machines of their own invention, will inevitably be forced to do what will ultimately become possible to do, namely, integrate their biology with computer technology. This result may conjure up images of the Borg on *Star Trek*, but I see no cause for worry. It is not clear to me that an intelligent semibiological machine is necessarily any better or worse, more or less emotional, or more or less moral than a human being. In any case, worrying about it is largely irrelevant. If it is possible, it will happen, as I expect cloning, genetically selective reproduction, and a host of other practices that have not yet even begun to give ethicists nightmares will also happen.

That is the *optimistic* outlook, from my perspective. Alternatively, there is the possibility that rampant population explosions, combined with ever scarcer resources on a hot, polluted planet, mixed with a possible victory of superstition and myth over logic and rationality, will result in numerous devastating wars, and perhaps the establishment of theocracies that suppress scientific thought, well before technological progress reaches the stage I have described. Human civilization then takes a giant step backward, forgoing learning, science, and rationality for an ordered subsistence living, forsaking the here for the hereafter.

Our future in this regard will affect the future of our oxygen atom. For example, because oxygen, via controlled combustion, is one of the chief motors of the engine of life, if we change the way the motor works the cycling of oxygen in the atmosphere will change.

Nevertheless, life in any form requires energy, and because sunlight is such an abundant free form of energy, I have a hard time imagining that some form of solar power, via photosynthesis or some other scheme, will not be utilized by intelligent life. Photosynthesis converts solar energy to help break apart water in order to gain hydrogen atoms and energetic electrons that can be used later to help build up organic molecules. But what if conscious life becomes more and more tied to silicon, and not carbon? The ultimate goal may no longer be to generate organic molecules capable of facilitating chemical reactions allowing reproduction to occur. Energy will still be needed to drive the processes in silicon, even if these primarily involve inorganic reactions. Light will still be free in the future, even if it may be more sporadic, and thus it seems reasonable to suspect that life will continue to exploit it.

By the same token, because oxidization effectively allows stored energy to be released I expect that oxygen itself will also continue to remain vital to the process of life, even if life may not depend on respiration as we now know it. Remember that what oxygen primarily does is simply take up spent electrons. If life becomes dependent on electric currents in semiconductor chips, something will still have to drive the currents, which means moving electrons from high energy to low energy, and something will have to facilitate the transformation, and take away the spent by-products. It will be intriguing to see whether intelligent life, as it designs intelligent life, may be able to improve upon random natural selection. Possibly. As my friend physicist Frank Wilczek has stressed to me, evolution never invented the wheel!

Finally, there comes the whole question of reproduction. We exist, as the biologist Richard Dawkins has so ably put it, merely as conveyers of genes. Apparently the purpose of life is simply to provide a robust mechanism for certain molecules to multiply. All of our hopes and dreams, lusts and madnesses, go along for the ride. If consciousness can be embedded in a self-programmable system, will its imperative involve reproduction, or merely repair and improvement?

This seems to me to be a fascinating question, because if rampant reproduction no longer serves as the overarching purpose of intelligent life, then the strategies of intelligent life, built up over eons of biological evolution, may also change. If reproduction diminishes in importance, will the demands on oxygen diminish?

As interesting as these questions may be, they are probably only of philosophical relevance. They may affect the detailed life cycle that transforms oxygen and carbon to carbon dioxide, and oxygen and hydrogen to water. But the gross features of oxygen consumption will, I expect, depend on far more powerful energetic considerations than intelligent life may directly bring to bear. With these issues, and caveats in mind, let us peer into the looking glass.

Over the next hundred years, the chief development affecting the yearly routine of our oxygen atom, barring a major nuclear war, biological terrorism, or similar catastrophe, is likely to be the slow but persistent buildup of carbon dioxide and other greenhouse gases in the Earth's atmosphere. The impact will no doubt be subtle compared to the much more dramatic developments like Snowball Earth that have taken place over million-year intervals in the past. In the first place, the biosphere will no doubt respond to human production of carbon dioxide in ways we cannot yet currently envisage. After all, the atmospheric budget of carbon dioxide is minuscule compared to the total reserves on Earth, and the ultimate ability of the ocean to dissolve carbon dioxide is still far from fully understood. Nevertheless, it is hard to imagine that some sort of buildup will not occur. In the process, weather is likely to become more dramatic, with greater swings of temperature and of precipitation. The latter will affect the rate of carbon dioxide fixing in rocks, and also the production of oxygen by photoplankton in the ocean.

But such a variation in conditions is likely to be transitory, given the long-term global carbon dioxide terrestrial feedback mechanisms. The impact, on a global scale, will probably be marginal, no matter how much it upsets the course of human civilization. Coastal cities may flood, millions may die, but life will go on, and the continents will continue to evolve.

The burning of fossil fuels will end sometime in the next millennium, I expect, either because of the environmental impacts, the availability of new energy sources, the demise of civilization, or the increasing difficulty of finding such fuels. In any case, our oxygen atom is likely to pass through the blood of your children's children, and maybe your children's children's computer friends for some time to come. We could continue to follow tens, or perhaps hundreds, of such generations and oxygen cycles, but these would still represent a speck in cosmic time, and we have much bigger fish to fry.

Let us assume, however, that technological civilization persists long enough into the future to begin the terraforming of Mars (i.e., making Mars habitable). This imperative will become important on a much longer timescale. It is again hard to imagine that a civilization with the energy needs of our own will not exhaust many of Earth's resources sometime in the not too distant cosmic future. But before these resources are exhausted, global warming, or pollution of one form or another, or even overpopulation will probably make much of the Earth an unpleasant place to be. If life always exploits all available resources, and the past 4-billion-year history of life on Earth suggests that it does, it will be natural to seek out the resources of our nearby planetary neighbors.

So let us assume that life, and some remnants of our intelligence, populates a number of different locations in our solar system over the next millennium. At the same time, once significant intelligence can be encoded on microscopic chips, it is also hard to imagine that we would not send out miniature spacecraft housing these mini-me's and you's on long-term, one-way journeys of exploration beyond the solar system. In this way, I will also assume that some remnants of our intelligence will expand beyond our solar system.

Were our oxygen atom taken aboard such spacecraft, either for inter- or extra-solar-system travel, as fuel for rockets or fuel for life, its future would be immediately affected. But that seems too melodramatic, so I am going to assume that our atom remains stuck on Earth for the medium term, even if we do not.

In this case, we should fast-forward. The future of life on Earth is now irrelevant. We will meet up with it somewhere in the cosmos later. What is the future of our atom? On a timescale of thousands of years, weather variations will produce ice ages in some places, and deserts in places that are now green. But it is unlikely that such weather variations will completely alter the oxygen–life connection.

I am going to assume that we somehow deflect the next one or two planet-killing asteroids that come our way over the next 100 million years. If we do not, then a wholesale extinction of species, combined with dramatic variations in climate over years, centuries, and millennia, will occur. But in due time, the gross features of the biosphere that have persisted for billions of years will re-emerge. We know that every 200 million years or so, continental drift causes the complete complexion of Earth's continental crust to alter. Continents as we know them will disappear, and new ones will rise up. If a supercontinent once again forms near the equator, it is conceivable that another period of global glaciation could occur. The longer we look into the future, however, the less likely this possibility becomes. Quite the opposite, in fact.

When we consider a timescale of 1 billion years into the future, we begin to confront the first among several extraterrestrial challenges for all life on Earth. One of these will surely kill everything off, but it is not clear which one will do it. After all, life has survived, against many odds, for perhaps 4 billion years thus far.

In any case, one thing that intelligence on Earth is likely to be powerless to affect is the evolution of the sun. Remember that in the early stages of the Earth's history, the sun was 30 percent less bright than it is now. In another 2 billion years, the sun will be about 40 percent brighter yet. Unless someone or something intervenes, this increase in solar brightness will be fatal.

We know more or less precisely what will happen under these conditions because we have a sister planet located closer to the sun than we are, which at the present time receives as much sunlight as we will within 2 billion years. This planet is Venus, which we have already visited in this regard, and where surface temperatures exceed 400 degrees Celsius.

In fact, this will be tepid compared to the actual temperatures that can be expected to be reached on Earth at this time. For Earth will suffer the same fate as Venus, a runaway greenhouse effect, but on Earth, because of the presence of vastly more water, the greenhouse effect will be even more dramatic. One might expect surface temperatures to exceed 1,200 degrees Celsius! This will be enough to actually melt some materials in the crust.

As I alluded to previously, a runaway greenhouse effect would occur once the temperature on Earth rises by more than about 10 percent of its present value. The atmosphere would be able to hold much more water vapor than it can at the present time, so that some of the ocean water would evaporate. But water vapor is a greenhouse gas, and this would absorb solar energy and store it in the atmosphere rather than re-radiating it into space. Thus an increase in water vapor would result in a further increase in temperature. But this increase would allow yet more water vapor to be held in the atmosphere, causing the oceans to evaporate some more, intensifying the greenhouse effect. This would raise the temperature yet again, and so on. In the end, all of the water in all of the oceans on Earth would evaporate, and the temperature would then stabilize at the value mentioned earlier, more than 1,000 degrees hotter than the present Earth's surface temperature.

No forms of life could survive these conditions. Some bacteria are hyperthermophilic, but these conditions would require hyper-hyperthermophilia. Such creatures could not be made of organic materials at all, as no such materials would remain intact at such absurdly high temperatures.

The runaway greenhouse effect would also dramatically alter the conditions for our oxygen atom on Earth. If all of the water on Earth evaporated, the resulting atmospheric pressure due to water vapor would be about 100 times as great as the present pressure. Water vapor would thus be, for some time, the dominant gas around. This density of water would not survive in the atmosphere for long, however. After all, when we measure the amount of water in the Venutian atmosphere its abundance corresponds to about 1/100 of the pressure of the present Earth's atmosphere. Originally, Venus must have had an abundance of water similar to Earth's, simply because it formed in a similar environment and is a similar mass. What happened to all Venus's original water?

Water vapor in the atmosphere can be broken apart by light from the sun. This happens with low efficiency, but it happens. When it does, the hydrogen gas, being light, will, at the temperatures we are now considering, easily evaporate off the planet almost completely. We know this has happened on Venus by comparing once again the abundance of normal hydrogen in the atmosphere of Venus with the abundance of its heavy cousin, deuterium. Since deuterium is twice as heavy as hydrogen, it will be much less subject to evaporation from the planet. The deuterium-to-hydrogen ratio on Venus is about 150 times greater than it is on Earth. This suggests that as much as 150 times as much hydrogen once existed on the planet as exists now. If all of this hydrogen came from water, there could have been several meters' worth of water covering the whole surface of Venus in early times.

In fact, we think that even more water than that once existed, perhaps an amount comparable to that on Earth. If we spread out the water in all the oceans today over the whole surface of the Earth, it would form a layer about 2 kilometers deep. This is almost 1,000 times as great as the abundance of water inferred to have existed on the primordial Venutian surface based on the above estimate. Where could we have gone wrong? (I say "we" here, even though I led you through the estimate, but of course, you didn't have to agree with me!) We were forgetting that comets and the like could have fed water to Venus throughout the age of the solar system, as they have undoubtedly been feeding the Earth. About 10 to 20 comet impacts would be all that was required to have supplied as much water as is now measured to be on Venus. But these impacts would have led to a deuterium-to-hydrogen ratio that is close to the terrestrial value. In order for the present Venutian value to be so much greater, substantially more water must have been present than an estimate that ignores the effect of comets would suggest. And thus substantially more water must have disappeared from its atmosphere. The same thing could occur over the course of time on the scalding planet Earth of the future.

As this disappearance occurs, more oxygen atoms will be left around with no hydrogen to bond to. Thus for some time it might be possible for oxygen to build up in the atmosphere. But with no oceans left, the carbon dioxide feedback cycle would now be broken. There would be no rain to scrub out carbon dioxide, no surface water to contain the carbonic acids, and thus no carbonate formation from rocks. The sink for carbon dioxide would thus disappear from the Earth. But the volcano source of carbon dioxide would persist because presumably plate tectonics would persist in some form. Thus carbon dioxide would continue to build up. The present Venutian atmosphere has as much carbon in carbon dioxide as Earth has carbon in any form, including in carbonate rocks and organic materials. Once the carbon cycle ends on Earth, and carbon is simply fed into the atmosphere without being scrubbed out again, eventually it will be possible that the Earth's atmospheric concentration of carbon dioxide could approach that of Venus. It is unlikely, however, that the oxygen fraction would remain significant in this atmosphere that is 100 times denser than our present atmosphere. As the oxygen abundance increases, more and more materials on the surface of the Earth will spontaneously combust, getting oxidized and producing quantities of carbon dioxide, silicon dioxide, and sulfur dioxide in the process.

In this future lifeless hellhole, our oxygen atom would have one of two fates. Either it would be stored in oxidized rock, or it would combine with carbon to form carbon dioxide. There would be no more life cycles, or even global carbon cycles, to speak of. Every day, for billions of years, the events of the day would be completely describable by a single word: *hot!*

Of course, like the Ghost of Christmas Future, I am merely presenting here the future of the Earth if we don't do anything about it. One possibility was suggested to me by the Dutch physicist Gerardus 't Hooft, an avid science fiction fan who likes to think outside the box. (He is good at it, too. He shared the Nobel Prize in physics in 1999 for having creatively solved one of the most important problems in particle physics while he was still a graduate student in 1972.) He suggested to me that a future technological civilization billions of years from now would work to move the planet out of its present solar orbit, so that it would be farther from the sun. In this way, it would not heat up as the sun's luminosity increased.

I don't think this will happen, however, even if humans have the technological resources to accomplish it. The reason is simply that if you work out where the Earth's orbit would have to be at this time in order for temperatures to remain conducive to life, you would have to move the Earth out to a distance not significantly closer to the sun than the present position of Mars. But why move a whole planet out to such a distance when a perfectly good planet already exists there?

It seems instead much easier to move some fraction of the population of Earth to Mars. If one could arrange for the substantial release of carbon dioxide from Martian rocks, one might hope to create a beneficial greenhouse effect on the planet that would warm it up and melt the water in its polar caps. Could we then import photosynthesizing plants to create oxygen, and slowly build up a home away from home? It is not clear, but we still have plenty of time to work out the details — unless that killer asteroid hits tomorrow, that is.

In any case, even such extreme measures will only temporarily stave off the inevitable. Within another 4 billion years, the Earth will, on its own, migrate out to the present position of Mars. But by then it will soon be too late for anything to survive in the inner solar system. By this point the sun will have reached twice its current brightness, but more important, all of the hydrogen fuel in the solar core will have been converted into helium. The solar core, not yet hot enough to fuse helium into yet heavier elements, will begin to collapse and heat up, releasing that heat to the regions outside the core. These will begin to burn hydrogen in a frenzy. At the same time, the sun will begin puffing up in size due to the heat deposited in its outer regions. Its surface will cool as it expands, so that the light emitted at its surface, at about 3,000 degrees Kelvin, will be redder than that emitted by the present sun, with a surface temperature of almost 6,000 degrees Kelvin.

Nevertheless, as cool as it will be, by the time the sun has puffed up to its full Red Giant glory its size will encompass the present orbit of the Earth, and its brightness will be more than 1,000 times that of the present sun. Imagine the entire sky filled with a ball of fire 1,000 times brighter than the noontime sun today!

The Earth, however, will no longer be in its present orbit. As the sun burns material in its outer regions, the engorged star will begin to eject particles from its surface. Perhaps a quarter of the entire mass of the sun, a mass equivalent to perhaps 100,000 Earths, will be thrown out into space over the course of a few hundred million years. The Earth, in turn, will begin to migrate outward in response to the reduced gravitational pull of this new leaner, meaner sun. It will settle into an orbit close to the present orbit of Mars.

It is hard to imagine any atmosphere on Earth being able to withstand this onslaught of radiation and solar flotsam and jetsam. Surface temperatures on the Earth will now be such that it may become molten, and the atmosphere will likely boil off, or be driven off by the strength of the solar wind.

Even if somehow, by some miracle, our oxygen atom withstands this barrage, one last final indignity will likely send it flying out into the cosmos. Within 1 billion years of exhausting its hydrogen fuel, the core of the sun will finally heat up to a temperature of about 100 million degrees. At this temperature, helium will begin to be fused into carbon. I have already described how, because this reaction depends sensitively on temperature and because of the rather peculiar configuration of matter in the solar core, this turn-on will be truly explosive. The entire core of the sun will briefly turn into one massive thermonuclear bomb! During the helium flash that results, the sun's core will produce as much energy as is produced by all the rest of the stars in our galaxy. Most of this energy, however, will be absorbed by the outer layers of the sun, which will expand even more. The Earth is now likely to be toast, and our oxygen atom will probably be streaming full speed ahead away from a home that has sheltered it for more than 10 billion years and out into the darkness of space.

Any remnants of the once great continents and oceans of the former blue-green planet will have been erased. Its surface will resolidify as the sun calms down and the Earth gains a new face, displaying no memorial to the countless generations of miraculous life-forms that once inhabited its surface. Any life that did not escape the solar system far earlier will simply be obliterated. There will be nothing, and no one, to take even one last breath . . .

18.
ASHES
TO ASHES

*Sunt lacrimae rerum et mentem
mortalia tangunt.*

*These are the tears of things, and the
stuff of our mortality cuts us to the
heart.*

VIRGIL

In 1880, Auguste Rodin was commissioned to produce the piece that occupied him, on and off, over the next 20 years: a set of great bronze doors for a proposed Musée des Arts Décoratifs. The theme he created for them, *The Gates of Hell*, based on Dante's *Inferno*, was to provide him with subjects for many of his most famous independent creations.

Cast in bronze a decade after his death, the doors now sit on display in an outdoor garden of the Musée Rodin in Paris. Over 200 individual figures decorate the lintels and the face of the doors, depicting a descent into hell by fallen souls whose strained bodies mix suffering with sensual desire.

In the center of the tympanum, the panel above the door, sits the prototype for Rodin's most famous sculpture, *The Thinker*. This was originally intended to represent Dante, seated in reflection and temporarily separated from Virgil, his guide through the netherworld. Instead it became an anonymous, powerfully muscular savant, forever embodying man's reflection on his fate. In Rodin's words, "Fertile thoughts slowly rise in his mind. He is not a dreamer. He is a creator."

In our atom's life story we have now crossed the threshold beyond the end of life on Earth. The planet has descended into its own hell and is now devoid of life or other distinguishing features. Have our dreams evaporated into space along with everything else? Or have our creations rescued us, so that some remnant of our consciousness persists somewhere out in the cosmos? Rodin's *Gates of Hell* will have by then long ago been destroyed, along with everything else on the Earth's surface. Yet a civilization that could produce such a masterpiece may have the creativity and voracity necessary to survive. But even if our descendants, in one form or another, do survive, how long can life keep postponing what seems to be the inevitable end? Can our civilization, if not our species, ever hope to mimic the apparent immortality of our atoms?

After all, this great demise is simply yet another new beginning for our atom. It has spent 10 billion years on a single planet, by far the longest period of staying put since it was created. Nevertheless, the largest part of its existence is still yet to be — so large a percentage, in fact, that I cannot even begin to give it proportionate weight here. I have already spent a couple of hundred pages describing its first 15 billion years or so. Imagine instead that I had spent only a single page on this entire interval (maybe you wish I had!). If every page of this book spanned another 15 billion years, there are not enough pages in all the books in all the libraries on Earth to equally span every further 15-billion-year interval to come.

Thankfully, however, we needn't do so. All terracentrism aside, it may be reasonable to claim that more transformations have taken place for our atom during its life on Earth than will occur in all the rest of future history, recorded and unrecorded.

Of course, this claim depends in part on whether our oxygen atom has the fortune to once again, at some point in its existence, become part of a living planet. The likelihood of this is not so easy to estimate, because we do not know how probable life is in the universe, after all. If life is a one-in-a-million proposition, it is virtually impossible that our atom will see any further action. If we are optimistic and assume 1 in 10 planets house life (as is at least the case in our solar system), the odds are much better, but it is still a long shot. After all, 99.9 percent of the mass of our solar system is contained inside the sun. If our atom's future is in another solar system like ours, there is thus likely to be only a 1 in 1,000 chance that it will not end up inside a star. Only objects with a mass of more than about 10 percent of our sun become stars, so even in a solar system with a smaller-mass star, and planets no larger than Jupiter, the probability is still at most 1 percent that our atom will end up outside the star. Indeed, our atom has already been a part of two planets, beating the odds. It is thus not likely to do so again. Depending on the star it falls into, that could be the last thing it ever does.

Even if it is blown out again, how many more such cycles is it likely to have? Star formation is certainly continuing with great profusion today, although probably not as fast as it did early on in the galaxy's life. One simply has to look up to the constellation of Orion, one of the easiest for me to find in the sky because it looks like a perfect stick man and thus matches exactly my own artistic abilities, to find a rich star-forming region today. When seen with the benefit of the Hubble space telescope, rich globs of shimmering gas appear, and several protostars with beautiful extrasolar disks can be discerned. We are looking back at circumstances that mimic our own birth!

At the current rate of star formation in our galaxy, the amount of free gas, combined with the amount of material stored in stars that will either explode or expel a significant fraction of their mass, will allow stars to continue to form for at least 10,000 billion years. This is a colossal amount of time, 1,000 times longer than the current age of the universe, and we cannot realistically begin to consider such timescales without taking into account possible changes in the overall evolution of the universe itself. If the universe were to stop expanding and were to recollapse, for example, it is likely that it would do so on a timescale much shorter than this.

But don't cancel your summer vacation yet. All current evidence suggests that the expansion of the universe will continue on indefinitely. Whether or not this is good news for life is something we shall discuss soon. Nevertheless, it does allow for the possibility that our galaxy will use up all its available stellar fuel before the universe ends.

A lot can happen before the galactic fuel gauge reads Empty, however. For example, about 5 billion or 6 billion years from now, around the time our sun begins its long Red Giant death march, it is likely that a period of suddenly enhanced stellar births and deaths will begin in our region of the universe. This is because we are currently on a collision course with our beautiful twin, the Andromeda galaxy. This nearest large galaxy similar to our own is about 2 million light-years away. Like Siamese twins, in fact, we seem to be irrevocably bound together. The Andromeda galaxy and the Milky Way are part of a local group of galaxies caught up together in each other's mutual gravitational attraction. Over the course of billions of years, these galaxies perform cosmic pirouettes around each other. Sometimes, like clumsy dancers, they may collide. The Andromeda galaxy is currently heading toward us at a speed of about 100 kilometers per second. It will thus cover the remaining 2-million-light-year gap in about 5 billion years.

The collision of our two galaxies will re-create, in slow motion, the collisions and mergers that put the Milky Way together in the past. This time it may assemble a new mega-galaxy. Nevertheless, the distance between stars is so great that, once again, actual stellar collisions will be rare. But over the course of perhaps 500 million years, the two galaxies will completely cross through each other. In the process, stars will be ripped from one galaxy and stolen by the other. Some stars will be ejected completely, flying out forever into the dark loneliness of intergalactic space.

While the net result will be dramatic, the disruption will be so gradual that were we in the midst of colliding with Andromeda now, the ongoing collision would have been completely unnoticeable throughout the whole of recorded human history. Indeed, only with the recent advances of astronomy, made possible by the large telescopes of this century, would we have been able to infer that such a collision was actually taking place.

But on a stellar timetable, such a collision nevertheless makes waves. Gas clouds throughout each galaxy are disrupted, and shock waves move throughout the galaxies. Whenever gas is compressed, stars form. To an observer who has a billion or so years to watch the process, when our two galaxies collide we will light up like a Roman candle.

The ultimate object, or objects, that result from the collision will probaby be dramatically different in shape from either the Milky Way or

Andromeda. But the burst of star formation that will occur during the collision is not likely to dramatically change either the amount of gas available in the long term or the total number of stars that should form before the system burns itself out. Thus we can optimistically assume that stars will continue to shine about 10,000 times longer than the current age of our universe, so we can make projections regarding how likely it is that our oxygen atom will end up on another life-filled world.

The simplest estimate would assume that the average star is about 1 solar mass, and 1-solar-mass stars live about 10 billion years, expelling about a quarter of their mass back into the solar system before they burn themselves out. Dividing the time left by this cycling time yields about 1,000 cycles. This would suggest that if our atom participates in all the cycles, it has a reasonable likelihood of ending up on a planet once again, although not necessarily one with life.

Unfortunately, however, this estimate is impossibly optimistic. In the first place, if three-quarters of the mass of 1-solar-mass stars remains within the stars, on average, then the probability of our atom successfully cycling in and out of even 10 stars in a row is less than 1 in 1 million. If it ever enters a star that does not explode, but dies with a whimper, whatever remains of the star will be likely to become our atom's last resting place.

But there is a much more significant reason that speculation based on the average is completely inappropriate. Smaller stars, called Red Dwarf stars, live longer than more massive stars. Such stars may be puny, with brightnesses less than 1/10,000 that of our sun. Nevertheless, these weaklings will one day inherit the universe. A star 1/10 the mass of our sun lives more than 100 times longer, and a Red Dwarf whose mass is at the lower limit at which nuclear burning can begin, or about 0.08 solar masses, can live more than 10 trillion years on its meager fuel supply. As stars continue to burn, and supernovae continue to explode, more and more hydrogen gas will be converted into heavier elements. As the heavy-element abundance increases, the lower limit on the mass of an object that can generate energy via fusion following gravitational collapse decreases to about half the present limit. As the lower limit on stellar sizes decreases, the lifetime of the smallest stars, some of which will have a surface temperature comparable to that of the Earth today, will also increase.

Thus as we wait longer and longer, a greater fraction of the gas in living stars will be in small, long-lived objects. Red Dwarf stars are already the most common type of star in our galaxy, although their heavier cousins still dominate in total mass. These bigger brutes, like many a foolish youth, will squander their lives by burning bright and dying young. Aesop taught us that slow and steady wins the race. Within about 100 trillion years, all the gas in the galaxy will have been used up, and the only objects still shining will be these faint red tortoises.

Actually, near the end of their lives, Red Dwarf stars briefly burn almost as brightly as the sun. Brief in the context of 10 trillion years means about 5 billion years or so. Thus if our atom were fortunate enough to find itself on a planet orbiting such a star near the end of the star's nuclear burning phase, there is enough time and energy for life to evolve and die on such a planet before the star does. The odds are incredibly stacked against this, however. Ultimately, before this is likely to occur, our atom, if it stays in our galaxy, will become a part of some other star and never emerge to bask in its light again.

This argument suggests that each atom in the universe is lucky to have one shot at life. Our oxygen atom has already had its go, and it will most probably never have another. But this does *not* mean that life itself forms only once in the universe. Rather, most atoms in the universe spend their time as part of systems other than living planets. They are lucky to visit such systems even once during their existence (although you may recall that at least part of our atom has been lucky enough to visit habitable planets twice, even if only one of them survived long enough for life to evolve). If this confuses you, think of the following analogy. Each of us may have, at best, one shot at becoming a millionaire in our lifetime, but this does not mean that there is only one millionaire on Earth!

What of the future of life in a slowly dying galaxy? We are greatly aided here by the certain fact that life exists at the present time. This greatly increases the a priori likelihood that it will be found somewhere in the future. When our Earth becomes a lifeless ball, perhaps 2 billion years from now, any descendants we have, electronic beasts or flesh-and-blood animals, will have either long ago departed on a pilgrimage to a new safe haven, or they will have long ago departed in a different sense.

If conscious beings do escape, they will presumably do so on large ark-like spaceships that are not designed so much to go somewhere as to go anywhere. At a minimum, tens or hundreds of thousands of years will be required to span the space between the stars. The spacecraft, like the solar system in which we live, will need to be self-sustaining environments that move throughout the galaxy. Constant searching may eventually yield a suitable planet to touch down on, but whether the nomadic inhabitants of these craft will want to do this is not so clear. In any case, unless some galactic tragedy far more traumatic than the upcoming collision between the Milky Way and Andromeda occurs — perhaps the collision of two collapsed stars, which may vaporize a fair fraction of the galaxy — one can at least imagine life continuing around stars for billions, if not tens of billions, of years to come.

There is, of course, another option. If we are too stupid to survive, and our DNA gets roasted along with that of all the other animals over which we claim to hold dominion on Earth, could we nevertheless unwittingly seed the rest of the galaxy? Some organic material will probably survive the shock waves that expel material out into the solar system before, or perhaps following, the helium flash from the sun, just as organic material has survived cataclysmic collisions of meteors and comets with Earth. Could this material, along with our oxygen atom, be distributed to the cosmos, one day perhaps to help sprinkle another world with the seeds of life?

Here we are as aided by statistics as our atom is constrained by them. Every probability that holds for our oxygen atom holds for all the other atoms that stream away from our dying sun. But our atom is alone against the universe, so that when we discuss its future, we know that very rare events will probably never happen. When we speak of *many* atoms, however, we know that rare events will *always* happen. While most of the organic material torn off of our planet will either not survive the trip or will end up in the core of a star, one particle in a million, or even a billion, will be likely to end up on planets around the stars. Some fraction of these planets may be conducive to life. And on these worlds, our DNA remnants may prove crucial.

I refuse to harp on the details beyond this. Speculation is fun, but in science it is most fruitful to limit the details of one's speculation to a level appropriate to the data available. I cannot claim to know the future of the human species into the next century, even though journalists never tire of asking me about it. Once we go a few billion years into the future, it seems ludicrous to speculate on specific details.

All the same, I wonder whether these future catastrophes will really be catastrophes after all. Perhaps this formulation simply displays the limits of a "linear" mindset. If I think back over the history of life on Earth, it is the catastrophes that have driven life, in general, forward, even if they may not have been kind to individual species. The oxygen catastrophe, during which this potentially toxic gas began to build up in the atmosphere, drove life to develop respiration, without which multicellular animals could never have flourished. The Cambrian explosion, in which modern multicellular life began to diversify and take off, was perhaps prompted by Snowball Earth and its aftermath. Maybe it is necessary for many species to die in order to allow ones with more potential to get a foothold. Indeed, last but not least in this long line of catastrophe-induced progress, the extinction of the dinosaurs 65 million years ago made the rule of mammals possible.

Dante emerged from the horrors of the underworld a stronger and wiser man. Will the unavoidable descent into hell that is to occur on Earth make our heirs stronger, or replace them with something that is?

Whatever the future brings for life, and I really am of two minds on this, I nevertheless do not like the vision I have thus far painted of our oxygen atom's future: spending what is left of eternity deep inside the core of a long-dead star. Indeed, in such a scenario it is not likely to end its days as oxygen at all. If it joins a collapsing molecular cloud destined to one day produce a supernova, it may be converted into a heavier element still, transforming to iron, or beyond. Or, if it becomes trapped deep inside a supernova core, within seconds after the explosion begins it will see the past tens of billions of years of its history reverse themselves, as it gets torn apart into its elementary constituents. Its protons will be barraged by electrons and will convert to neutrons, which will merge with the rest of the stellar core into a neutron star. Or worse still, the star it will become a part of could be so massive that its collapse will not be halted by a nuclear bounce. Instead, our atom, and the rest of the stellar core, will continue to fall inward, forming a black hole. Once our atom crosses the boundary where not even light can escape, it is lost forever from our universe. Its fate is not likely to be pleasant, but we will never know what hits it.

The realization that I am the author of our atom's fate empowers me. I can choose a different future, and I will:

Our atom has ended its age of innocence. It has witnessed the birth and death of planets, and more species of life than we could ever name. Thrust out, away from the dying solar system, it travels outward for several million years, and its next home is in a molecular cloud near the outer edge of the Milky Way. This cloud collapses to a star that will one day undergo a supernova explosion. But our atom's nucleus (the electrons will be stripped off soon after the star forms) will remain again outside the densest part of the core. It may momentarily assist in the formation of helium by absorbing a proton, but it will quickly give it up again, preserving its identity over another 100 million years until the star goes *boom*. But this time, the neutron star that forms will be spinning rapidly. Huge magnetic fields will steer charged particles outward in a jet of matter emanating near the axes of the rotating star. Our oxygen nucleus will get shot out along this jet, being further accelerated by the shock wave that it passes through. In the end, it will be traveling outward at nearly the speed of light.

In this way, our oxygen nucleus will leave the galaxy, traveling outward through space, out of our local group, and eventually out of the cluster of galaxies that contains it. It is now free to travel throughout the mostly empty universe unhindered. The likelihood that it will ever encounter another atom gets smaller and smaller over time, as the universe expands and the average density of matter diminishes. Its immediate future now depends on the future of the universe as a whole, and not on the capriciousness of its constituents.

You needn't think that my choice of our atom's future is particularly anomalous or whimsical. Existence outside the confines of a galaxy is in fact far more common than the alternative. The bulk of matter in the universe, at least twice as much as can be accounted for by all the mass in stars and galactic gas, appears to exist in the form of hot gas outside of galaxies. It is likely that much of this gas was literally blown out of the galaxies by supernovae or other energetic events. Our atom merely gets slightly more energy than the norm, so that it no longer remains confined by even the gravitational attraction of the largest known conglomerations in the universe, clusters of hundreds of galaxies spanning tens of millions of light-years across. But this is also not unusual. We are bombarded daily by cosmic rays coming from distant galaxies, among which are included the nuclei of heavy atoms thought to have been expelled originally by supernova events.

Our atom may now travel for billions and billions of years through the vast emptiness of space, heading nowhere in particular. Moreover, if it is traveling at close to the speed of light, its internal clock will slow down compared to local time in the galaxies it passes by. Billions of years may seem like hundreds of thousands of years, which is, for our atom, a very short span of time. It may cross our currently visible universe in what will seem like less time than it spent being part of the sweat and blood of humans and their ancestors. In this way it could live out an eternity, alone, but free.

But while it may escape the grasp of galaxies, our atom cannot escape the inevitable consequences of the laws of physics. Oblivious to its fate, it is nevertheless heading toward an end that was engraved the instant its protons were first formed in the heady microseconds at the birth of our universe.

19.
DUST
TO DUST

Eternity is a long time, especially toward the end.

WOODY ALLEN

Some years ago a memorable commencement exercise took place at a well-known private girls' school. The speaker is reported to have proceeded to the podium, all the while surveying the smiling girls in their bright white graduation gowns, surrounded by their parents and friends. Looking out at the crowd, he began, in a booming voice: "Things are going to get unimaginably worse, and they are never, ever going to get better again!"

I heard this story as a student, and do not know whether it is apocryphal. Nevertheless, I have always wanted to recount it in writing, and this seems like the perfect place. As the universe evolves there will continue to be, at least metaphorically, moments in the sun, perhaps for all eternity. But these will be fewer and farther between. If we peer as far as we can into the future of Life, the Universe, and Everything, things do not look too good.

Discoveries made over the past five years suggest that we actually may live in the worst of all possible universes, at least in terms of long-term quality and perdurability of life. And the same theoretical developments that hold out such hope for understanding why we live in a universe full of matter also imply that ultimately, this matter is ephemeral.

Before 1916 it made no real sense to discuss the future of the whole universe, because no theory existed that could consistently describe the dynamics of space, time, and matter. In that year, however, Albert Einstein completed his greatest piece of work, the discovery that would make him a household name, and the Man of the Century. Against all intuition, the space in which we live is actually curved, and it is the matter and energy content of the universe that is responsible for this curvature. Einstein's general theory of relativity yielded the necessary links between space, time, and matter. As such, he quickly realized, it offered the possibility of not merely describing the motion of objects within the universe, but the behavior of the universe itself.

There was a problem, however. Einstein's universe shared with Newton's a fundamental problem: Gravity sucks! That is to say, gravity appears to be universally attractive, unlike, say electric or magnetic forces, which can be attractive or repulsive. As a result, there exists no stable configuration of matter spread throughout the universe. The gravitational attraction of massive objects will invariably cause them to collapse together.

Einstein's problem was that in 1916, conventional wisdom held that the universe was static. After all, if we look up into the heavens, do not the distant stars appear immovable? Einstein was disheartened by this realization, but not daunted. He recognized that he could make a minor alteration in the equations governing the evolution of the universe that would solve this problem. He added an extra term, which he called the *cosmological term,* that would produce a very small *repulsive* force throughout all of space. On human scales, and the scale of our solar system, this small additional force would be unnoticeable. On the grand scale of the universe, however, this force could balance the gravitational attraction of distant galaxies and keep them apart.

Almost immediately, however, this idea ran into problems — so many, in fact, that Einstein quickly labeled it his "Biggest Blunder." We should all be so lucky.

First of all, within a decade or so of Einstein's "Blunder" it had become clearly established that the universe was not, in fact, static. By 1929 the American lawyer-turned-astronomer Edwin Hubble had published his study of an expanding universe. But even earlier, the notion that the universe might not be static was in the air. In 1923, Einstein wrote in a letter to the mathematician Hermann Weyl: "If there is no quasi-static world, then away with the cosmological term!"

You see, if the universe is expanding, there is no need for a universal repulsive force in nature. Gravity can simply work to slow the expansion. Depending on how fast the expansion is, and how much matter there is to slow it, the universe will either go on expanding forever or it will slow down, stop, and collapse. Trying to determine which future is ours then became one of the central industries of cosmological research.

Much as Einstein wanted to get rid of the now unnecessary extra term in his equations, this was a little like trying to get the toothpaste back in the tube. Had Einstein not introduced this term, someone else would have.

The cosmological term turns out to have a real physical significance, or at least a significance that may be real. In general relativity, energy, in any form, is the source of gravity and hence of the curvature of space. And it turns out that one very particular type of energy produces precisely the effect of Einstein's cosmological term. This is the energy of nothing at all.

Now, you may say that in any sensible world "nothingness" cannot be possessed of energy. But no one has ever claimed that quantum mechanics is sensible. In fact, quantum mechanics, when coupled with Einstein's special relativity theory, implies that empty space is not really empty. As I have described, it is full of a seething, bubbling brew of elementary particles, called *virtual particles*, that can suddenly appear and disappear back into nothing in a time so short you cannot observe them directly.

This may sound suspiciously like describing how many angels can dance on the head of a pin, but there is an important difference. Angels are invisible, and normally unmeasurable. Virtual particles are invisible, but they measurably affect almost every microscopic process in the universe. On such scales, for example, virtual particles can change the properties of atoms, by momentarily altering the distribution of electric charge. This in turn will affect the energy levels of electrons orbiting the atoms in ways that can be calculated, and the predicted change agrees with the measured change more accurately than any other prediction in all of physics.

If empty space is truly full of virtual particles, the question arises whether these particles can also contribute energy to space. If they can, then the cosmological constant will not be zero. Unfortunately, when we try to estimate how much energy such quantum effects can contribute, the typical prediction is about 120 orders of magnitude larger than all of the energy contained in all of the stars, gas, planets, and people in the entire visible universe.

There is a simple experiment anyone can do to demonstrate that empty space cannot contain such a mammoth store of energy: Look at your own nose! Indeed, as the novelist George Orwell once wrote: "To see what is in front of one's own nose requires a constant struggle." This takes on a whole new meaning in a universe with a cosmological constant. In such a universe, the repulsive force in empty space causes distant objects to move apart with a relative velocity proportional to the distance separating them. This means that objects separated by greater than a certain distance will actually be moving apart faster than the speed of light!

This may sound impossible, but while special relativity precludes objects traveling through space faster than the speed of light, general relativity implies that space — whose expansion can carry the objects apart — has no such restrictions. The effect of objects separating faster than the speed of light is that one object becomes invisible to the other, since the light from one cannot compete with the expansion of space in order to traverse the distance between them.

As a result, we can put a limit on the magnitude of a cosmological constant. If it were larger than a certain value, the space between your eyes and the end of your nose would be expanding faster than the speed of light. The light from the end of your nose would thus never reach your eyes. For this not to occur, the maximum allowed cosmological constant is at least 70 orders of magnitude smaller than the naïve prediction resulting from the estimate of the energy of empty space I alluded to above.

But we can do far better than this. We can see not only the ends of our noses, but, with our telescopes, throughout the universe. The fact that we can see distant galaxies billions of light-years away from us puts by far the strongest available constraint on the magnitude of a cosmological constant, about 120 orders of magnitude smaller than the naïve estimate above.

Nevertheless, as small as this value is, it allows for an absurdly small cosmological constant to exist. It is even possible that the energy density of space could still exceed that associated with all matter in the universe.

As absurd as this may sound, about five years ago Michael Turner, of the University of Chicago, and I, and also several other independent groups, proposed that this was the case in our universe. We recognized that the available data from cosmology, pertaining to the age of the universe, its composition, and the distribution of matter, when combined with theoretical arguments yielded a highly suggestive, and highly heretical, possibility: At least 60 percent of the energy of the observable universe seemed to reside in empty space.

If this were true, it would imply that the present expansion of the universe should be accelerating, rather than slowing down, as would be the case in all conventional theories in which the universe was dominated by matter or radiation. As if on cue, within two years two different international collaborations, one led by Brian Schmidt at Mount Stromlo Observatory in Australia and Robert Kirshner at Harvard University, the other by Saul Perlmutter at Lawrence Berkeley National Laboratory, reported the results of a bold new effort to measure the acceleration or deceleration of the universe.

They used as a guide a special kind of exploding star, called a Type 1a supernova. Unlike the exploding stars our atom visited prior to its time on Earth, a Type 1a supernova is thought to involve what is closer to a cosmic thermonuclear bomb. In this case, a star like our sun that has already completed its lifetime of nuclear burning ultimately passes through a Red Giant phase to a dense White Dwarf configuration. If this star accretes matter onto its surface from some nearby object, however, eventually the mass of the star will increase to the point where the pressure and temperature in its core are great enough to begin nuclear burning of helium and heavier nuclei. The sudden energy released blows the star apart.

The laws of physics imply a narrow mass range at which such an explosion can be initiated. It was thus suspected that such supernovae, which are so bright they can be observed across the visible universe, might provide *standard candles* whose intrinsic brightness could be a beacon by which we could measure the distance to faraway galaxies. Subsequent observations over the course of a decade seemed to establish a clear relation between the intrinsic brightness of such supernovae and the length of time they remain visible. By using this relation, one could hope to observe the brightness profile of yet more distant supernovae and with it estimate their distance from Earth. At the same time, we can measure the velocity at which the galaxies they are housed in are moving away from us. By seeing how this velocity varies with the inferred distance away from us, and recognizing that when we observe more-distant galaxies we are observing them as they were at ever earlier times (because of the finite speed of light), we can, in this way, determine how the expansion rate of the universe has varied with time.

There is one catch, however. Type 1a supernovae occur at a rate of only about 1 per 100 years per galaxy. But once again, we can make use of the remarkable fact that in a huge universe, even rare events occur frequently. If one manages to monitor about 30,000 galaxies in a single night of observing, statistics suggest that one should see 1 supernova per night. Using this approach, and the wide-field-of-view cameras that allow them to capture this many galaxies and more in a single image, both collaborations were able to discover multiple new supernovae in distant galaxies in a few nights of observing. By following the subsequent brightness profile over time, they could then estimate the distance to the galaxies, and hence the evolution of the expansion rate of the universe over time.

In January 1998 the groups independently announced a discovery that rocked the scientific community, and was even labeled the "Discovery of the Year" by *Science* magazine. Their data implied that the expansion of the universe is accelerating. What is more, the estimate of the required energy in empty space agrees precisely with that which we earlier inferred on the basis of indirect arguments. Now, in spite of this remarkable consensus, the supernova data remain preliminary, and it is premature to suppose that they will not later be proved wrong. But if they are confirmed, empty space possesses more energy, on average, than anything else in the universe!

If these observations survive the test of time, the implications for our understanding of fundamental physics are indeed profound. But the implications for the possible future of the universe, and of our atom, are even more dramatic.

If the expansion of the universe is accelerating, we can ask how long it will take before most of it has disappeared from our view altogether.

The answer my colleague Glenn Starkman and I recently derived is "Surprisingly soon," at least in a cosmic sense. Within about 150 billion years, all stars outside our local supercluster of galaxies will be traveling away at such a great speed that their light will no longer be visible to the naked eye. Within about 2 trillion years, not a single object outside our local supercluster will be detectable by any means whatsoever. We will become truly alone in the universe, forever and ever.

Of course, there is a bright side to every dark cloud. I have used this argument, for example, to emphasize to those in government who fund cosmology that if the expansion of the universe is accelerating, we have only a limited time to make our observations, and they had thus better give us the money now. More relevant to the matter at hand, once we acknowledge the possibility that empty space can have energy our ability to unambiguously predict the future of the universe goes out the window. Geometry is no longer destiny.

Without a cosmological constant, it was safe to claim that if there is sufficient mass around, this would ultimately slow the expansion of the universe, causing it to stop and reverse. If the mass density were not sufficient, the universe could expand forever. If vacuum energy dominates, however, then no matter how big the mass density, and how this matter density might alter the geometry of the universe, it cannot be sufficient to stop an already accelerating universe. Alternatively, even if the mass density is too small for its cumulative gravitational attraction to slow the expanding universe, a negative energy in empty space could still result in an extra-attractive force throughout space. This would eventually stop the current expansion.

Michael Turner and I have argued, in fact, that no finite set of measurements one can make over any finite amount of time will allow us to know unambiguously the ultimate fate of the universe. As long as physics remains an empirical science, guided by experiment and observation, the ultimate future of the universe will remain an ultimate mystery.

Must we then end this tale in such a state of ambiguity? Is there nothing at all we can say about either the ultimate fate of life, or the ultimate future of our atom in such an uncertain universe?

Humans are innately optimistic about the future, perhaps because

without that optimism, much of the hard work of living may seem pointless. We believe our progeny will survive. Death and taxes, pain and suffering might never disappear, but somewhere our children will carry on.
If necessary, we will colonize the galaxy just as life has thus far colonized
every possible niche on Earth. Or so we hope.

Charles Darwin was just such an optimist. In the conclusion to *Origin of Species*, he wrote: "As all the living forms of life are the lineal descendants of those which lived before the Cambrian epoch, we may feel
certain that the ordinary succession by generation has never once been
broken. . . . Hence we may look with some confidence to a secure future
of great length."

Darwin undoubtedly thought life could survive the cataclysms that
will inevitably follow in Earth's future. But what about the eternity of
time that may exist beyond the demise of the Earth? In Darwin's day, before the expansion of the universe was known, scientists worried that
eventually life was doomed to what is known as "heat death." Eventually,
a static universe would reach thermal equilibrium, in which the whole
cosmos would come to a common temperature. If there are no energy
sources or sinks, the famous second law of thermodynamics implies that
in such a universe useful work will no longer be possible.

But life demands, as I have emphasized before, a localized departure
from global equilibrium. Life is a thief of energy, hoarding it for later
use. Eventually life spends its energy balance in the process of living,
and as required by the laws of thermodynamics it releases energy as heat
in the process. Heat energy can be used for work only by transferring
heat from a hotter object to a colder one. Once all there is in the universe is heat energy at a uniform temperature, there will be nothing left
to steal.

The discovery in the 1920s that the universe is expanding changed
everything, however. In an expanding universe the background temperature generally decreases continually, delaying continually the onset of
such heat death. A continually expanding universe that cools forever
thus appears to offer new hope for life.

There is a long road from hope to accomplishment, however. In practice one must always confront this Universal Fact of Life: Batteries are required. Life will always depend on inventing new ways to steal energy

from a changing environment in order to power the processes of reproduction and metabolism.

When life first arose on Earth, this larceny was probably accomplished by accident. With energy abundant, there was more than
enough opportunity for complex molecules to arise. But just as we will
one day in the not too distant future have to face the fact that the fossil
fuel energy reserves on Earth are limited, as time marches on all local
energy reserves will also dwindle. The march of time is a march toward
equilibrium, but if life is to continue, it must stave off the arrival as long
as possible.

On Earth, we know precisely how much energy is required to power
life, and if we continue at our current rate of energy consumption it is
the work of a moment to estimate how much time we have left. But estimates based on our current metabolism do not necessarily provide an
accurate guide for the future. If intelligence survives the solar system catastrophe and moves out to populate other parts of the galaxy, and if it
survives the collision of galaxies and the dying of stars, it will do so by
subsisting on ever less and less energy over time. There is only one way
to reduce its energy needs. Like many a civilization in those early *Star
Trek* episodes, it must ultimately shed its cumbersome body, seeking out
means of preserving itself at a lower energy cost.

As the stars dim, and the universe cools, civilizations will face the ultimate limits to their existence. Whether or not they can survive the challenge, even in principle, is not yet resolved. In a landmark paper
published in 1979, the brilliant British-born mathematical physicist
Freeman Dyson, of the Institute for Advanced Study in Princeton, laid
the groundwork for the challenge facing civilization in the Ultimate
Future.

It may seem that in a possibly infinite universe that may expand forever, there should be infinite energy reserves to draw on. This is probably not the case, however. As the universe expands, the density of matter,
and available energy, decreases. Life would be required to mine a larger
and larger volume to extract smaller and smaller amounts of energy. Indeed, my colleague Glenn Starkman and I have claimed that there is no
mechanism to extract an ever increasing amount of energy out of an expanding universe, even in an infinite amount of time. If we are correct,

then life must ultimately face a budget problem unlike any it has faced thus far: how to make a finite energy reserve last an indefinitely long time.

In his earlier work, Dyson addressed this very question. In a manner typical of his great ingenuity, Dyson, who freely admits to an eternal optimism, demonstrated that contrary to one's naïve expectations, eternal life with finite energy is not in principle impossible. He made two simple assumptions, which may not be realizable in practice but which are certainly at the very least plausible. First, assume that life can continue to modify its bodily housing so that its metabolism uses less and less energy with time. Second, assume that for a civilization of conscious beings eternal existence is equivalent to continuing to have conscious thoughts, so that an infinite number of conscious awakenings is equivalent to eternal life, if not for a single being, then for a civilization of beings. One can argue with either of these assumptions, but one must agree that this is a "minimalist" interpretation of conscious life. If life cannot achieve even this level of existence, then surely anything more is impossible.

In this regard, the physics of infinity is no less tricky than the mathematics of infinity. Dyson pointed out that an infinite conscious future is possible even if living systems remain unconscious for an ever-increasing amount of time. Like bears in the winter, life can hibernate. Even if the periods of hibernation get longer and longer, in a universe that persists forever there is an infinite amount of time to spare. This may sound like an absurdly baroque argument, but Dyson proved that in a universe with limited energy resources, there is no other possibility for life, even in principle.

But Starkman and I have recently argued that even this possibility is overly optimistic. We have claimed that even in an eternally expanding universe, life cannot persist forever. We base our claim on the assumption that ultimately, as energy reserves decrease, the laws of quantum mechanics will govern the future of life. One of the reasons the quantum world seems so strange to us is that with every breath we take, we deal in huge numbers of particles. We have seen how this fact can lead to remarkable connections between all living things. However, it masks the fact that we now understand that at a microscopic level, energy is transferred between objects in discrete amounts, first called *quanta* by the German physicist Max Planck. Once one begins to investigate the universe at a level where individual quanta become significant, its weirdness becomes manifest.

If one takes the discreteness of nature at a fundamental scale into account, then the mathematics of living is different from what it would be if one could continue to treat energy as a continuously variable quantity. We have argued that once the energy required to power a living system becomes small enough, then the discreteness of energy transfers will become important, and in this case life, at least life that continues to think new thoughts, cannot endure forever.

Dyson has likened the difference between our form of "quantum life" and his continuous version of "classical life" as the difference between whether life is ultimately "digital" or "analog." In the former case, like a programmable computer, intelligent computations can be discretized into a series of "bits," ones and zeros. One may use such ones and zeros to play chess, or even encode and play back complicated music. Analog systems, on the other hand, do not rely on such discreteness. Old-fashioned record players, for example, reproduced music by continually varying the pressure on a mechanical needle, which in turn converted these mechanical impulses into a continually varying electrical signal.

Dyson has accepted our conclusions about the ultimate limits to quantum living, but he has held out the possibility that life might ultimately avoid this quantum catastrophe if it plays its cards right. The future, one might claim, thus lies not in CDs, but in LPs! As an example, he has resurrected a famous bit of science fiction from the astrophysicist–science writer Fred Hoyle, a classic tale about a "Black Cloud," in which a diffuse cloud of particles roamed the universe, and all the while was actually a conscious life-form. Its thoughts and actions were encoded in the motion of the particles making up the cloud.

Such a black cloud, Dyson has argued, is an example of an ultimate classical life-form, one that, as it slowly expands, can have a metabolism which continues to utilize less and less energy without approaching the quantum limit, and one which can, in principle, sleep for extended periods.

This issue is not yet settled. Dyson, Starkman, and I continue our friendly debate about a subject that is perhaps of only academic interest at present. The energy resources in our galaxy may be finite, but they are immense. Even at our present rates of energy usage, life could go on living for an unfathomably long period, literally billions and billions and billions and billions of times longer than the present age of the universe. Yet I nevertheless find it seductive to imagine that arguments based on simple laws of physics might allow us, rudimentary beings that exist in the earliest energetic heyday of our universe, to peer into our looking glass and divine the remote future of civilization.

The debate I have outlined has dealt with the best of all possible worlds, one in which life is sufficiently intelligent and resourceful to avoid all of the practical pitfalls of living. Nevertheless, the real world may be much closer to the worst of all possible worlds. In fact, if there is a cosmological constant that ultimately governs the future evolution of the universe, then life is truly doomed. One can show that such a universe will quickly, at least quickly on a timescale of the arguments I have just described, be driven toward heat death. The universe will reach a constant temperature, and no prolonged useful work will be possible any longer. Of course, in such a universe, if it persists indefinitely, local fluctuations will always develop. These local fluctuations might be sufficient to cause life to once again arise in some form or another. But it will always be doomed to die out again. In such a universe, life may always exist somewhere, but its appearance must be flecting.

It is, of course, always possible that in such a universe, new phenomena will allow us to escape what now appears to be the inevitable future. Perhaps the exotic physics of quantum gravity will let us create new, baby universes, into which we can insert some remnants of our consciousness before they disappear down into a black hole. But such speculations are just that. For now, they remain merely the fodder for a science fiction movie.

So much for life. What about our atom, the true hero of this story? I began this tale with the notion that our atoms, not us, might truly have a taste of eternity. Certainly the past history of our oxygen atom makes all of human history seem insignificant.

But just as life may die before our sun does, the days of our atom may be numbered as well. I have argued that it may proceed unscathed out of our galaxy, and out of our cluster of galaxies, in what seems an eternal voyage into the darkness. But inside the protons and neutrons that make up the heart of our atom may lie a clock that has been ticking for more than 10 billion years, waiting for a signal embedded at the beginning of time itself. Upon that signal, as surely as our sun, and our galaxy, will end their existence, so will our atom.

We return to where we began, to the clear water inside a tunnel inside a mountain in Japan. The story I told in the first chapters of this book is the story of a remarkable accident, whereby a universe without matter suddenly became a universe full of matter. But if this is the case, the very processes that created the matter that makes up the universe of our experience will one day slowly return our dust to nothingness. I have argued that it is a departure from equilibrium that makes the universe interesting. Without such a departure, nothing of note would ever happen. But by the same token, the approach to equilibrium is unavoidable. Life may die off as a result. So too may matter.

If matter can literally arise from nothing, in this case a primordial sea of matter and antimatter that would otherwise have been destined to annihilate to radiation, then matter is destined to return to nothing as well. The inexorable laws of physics tell us that in this case the energy stored in protons and neutrons is as temporary as the energy stored in life.

The gargantuan Super-Kamiokande detector, or its future descendant, posited to be 10 times bigger yet and now in the planning stages, holds the key to the mystery of our atom's ultimate future. One day, during my lifetime, or perhaps the lifetime of my daughter, one of these detectors may yield the unambiguous signal of a single proton ending its existence. As I stated when we embarked on our voyage, there are over 10^{34} protons in the Super-Kamiokande detector. That none of them has yet decayed in its two years of operation tells us that protons, on average, live considerably longer than 10^{33} years, an eternity by present standards.

However, I also remind you that indirect evidence from the physics of elementary particles suggests that we are tantalizingly close to seeing a proton decay. Estimates for the lifetime of protons, within the currently favored fundamental models, are in the range of 10^{34} to 10^{35} years. These may of course change, but if these models are correct as they now stand, a detector 10 times as large as Super-Kamiokande should then be able to record the decay of a single proton in a year of continuous running.

This would be a momentous day in the history of science. Not only would such an observation confirm our notions about the ultimate unity of forces in nature, it would allow us to empirically test our understanding of the origin of all matter in the universe. It is one of the wonders of science that keep me going on the very bad days, that an experiment with glass tubes and a huge tank of water could yield a signal that can take us directly back almost to the beginning of time.

But as wondrous as this capability may be, such an observation will provide definitive evidence that these days of our atom are also numbered. Our atom may continue its cosmic voyage throughout the universe for what seems like an eternity. All memory of the star that sheltered the planet that housed it for a brief 10 billion years will have long disappeared. The memory of the galaxy that housed that sun will have long disappeared. Even the light from all of the stars in the universe may have long disappeared. Our atom will be truly alone in the universe. No one may be around to witness its last moments. Eventually, on a timescale a billion billion billion times longer than the lifetime of nuclear burning in the longest-lived star in our universe, after a host of civilizations may have come and gone, one day a single proton in our oxygen atom will go *poof.* Then perhaps a billion billion billion billion years later, the second proton will die. The process will continue until our atom, and all atoms in the universe, are no longer. The lives of our atom will have finally ended.

An accident of nature, 12 billion years ago, is likely to have led to a slight imperfection in the universe, a small departure from equilibrium. This resulted in the existence of matter, and ultimately of atoms in our universe. This imperfection is likely to be repaired 10^{35} years into the future. At stake will be the future of matter. But perhaps even after the demise of protons and neutrons, all may not be lost. If protons and neutrons cease to exist, they may in turn decay into electrons and their antiparticle partners, positrons. By this time, the universe will be too diffuse for electrons and positrons to find each other in the desert of largely empty space. Perhaps electrons and positrons do not decay. Can such a universe with perhaps only a single electron in each region that now encompasses billions of galaxies remain vital? It seems hard to imagine. But the legacy of science has been that the universe is not limited by our imagination.

There may be no ultimate purpose to our existence or the existence of our atoms. The universe may become unimaginably worse, or it may not. There may be no reward in heaven. But surely the possibility that we, as conscious beings, have some hope of unraveling the secrets of a mysterious universe in the time we have allotted is itself a precious gift we should not squander.

Our atoms are vibrant messengers from the past, and harbingers of the future. They connect us in a definite way to everything we can see about us. Let us enjoy, with them, our moment in the sun.

EPILOGUE

I am drawn to the ocean. As I walk along the beach, the waves are soothing and the sun is warm. I know that in these very waters long ago, form arose from formlessness. These past two years I have lived and breathed through my atomic friend, and while I knew, or thought I knew, what this story was about, I wasn't really prepared for all the places it would lead me.

It still seems almost surreal to imagine that so much could have happened to make the simple act of my standing here possible. Can each of the atoms in the air I breathe really have gone through hell and back, braved the bitter cold of space, the brutal heat of stars, have crashed into the Earth, have dredged down below the continents and ocean floor merely to rise again? Have these atoms been a part of countless lives, and seen countless deaths? Will they travel throughout the cosmos that I would give my eyeteeth to explore?

I return, in my imagination, once again to my museum in Paris, to walk among my old friends, the seemingly eternal products of Rodin's imagination. Today I am taken not so much by the transformation of stone to skin as by the realization that water too is there everywhere: Adam and Eve in the water, Paolo and Francesca engulfed in an eternal brace amid the waves.

I walk into the waves in front of me. I dive in, for a moment not knowing if I shall ever resurface. In this split second, I realize that it doesn't really matter. Whether or not I survive, I know my atoms are likely to return to the ocean depths. What happens to them after I cease to exist is beyond my control, and their future seems inevitably written, regardless of my own hopes and dreams. I am only a temporary abode, and my life is an inconsequential moment in their vast eternity.

Yet I do surface. I am not overwhelmed by a sense of futility. I burst above the waves to take in a deep breath because I know that each breath takes me deeper into a great mystery story, and I cannot put it down. The universe remains full of such mysteries, from the origin of life to the ultimate future of our cosmos. They beckon us onward.

I have always taken solace from the myth of Sisyphus, who rolled his giant boulder toward the top of a mountain only to be doomed for all eternity to have it fall back down to the bottom. The odyssey of our atom may teach us that catastrophes can breed hope, and that one may never really know what is just around the corner. New wondrous experiences may await that can more than justify the pain of taking the next step. Like Camus, I have always believed that Sisyphus was smiling.

SOURCES AND
ACKNOWLEDGMENTS

O ne of the joys, and trials, of writing is the experience of learning how much one hadn't known about various subjects in advance. Nevertheless, with the exception of *The Physics of Star Trek*, which required research involving numerous all-night video sessions, the rest of my books have been centered on material about which I presumed in advance to have some professional expertise.

Even in these instances I found writing each book to be a remarkable learning experience, which is perhaps one of the reasons I continue to be drawn to writing. In any case, I knew when I decided to take up the challenge of the present work that the experience would be qualitatively different from anything I had done before. The prospect of using the lives of an atom to present a literary view of the history of the universe, including the romance of our own human drama, became more seductive the more I thought about it. It was also clear that this story would involve not merely physics and astrophysics, but at the very least geophysics, geology, astronomy, biology, and paleontology. Frankly, initially this challenge was also an attraction. I felt that were I to focus purely on physics in a new book, it would be too tempting to fall into the trap of repeating myself. In addition, I wanted the opportunity to address a wider audience, and I felt the breadth of the subject matter should match my ambitions.

As the scale of the task I had assumed became clearer, however, the challenge I had ahead of me also came into sharper focus. I knew the broad-brush outlines of what I would want to talk about, but in order to tell much of the story, I first had to assimilate many of the details myself. Over the past two years, I have had the benefit of tapping the accumulated wisdom and expertise of a host of individuals, a number of whom I would like to explicitly acknowledge here.

The areas of geology and biology represented two of the more obvious gaps in my education, and to begin to fill them I turned to academic colleagues with whom I felt comfortable enough to ask stupid questions. Ralph Harvey and Sam Savin of the geology department at Case Western Reserve University were kind enough to have extended lunches with me during which I bombarded them with questions about chemical evolution on Earth. In addition, serendipitously, over the course of several years of speaking engagements associated with *The Physics of Star Trek*, I had managed to make the acquaintance of a variety of experts in areas of key importance for this book. First, I thank Lisa Stubbs of the Lawrence Livermore National Laboratory for providing me, during meals and breaks over the course of a several-day meeting organized by the U.S. Department of Energy, with a primer on modern genetics. At another event, in Washington, I had the great fortune to meet Andy Knoll, of the Department of Organismic and Evolutionary Biology at Harvard. Andy is not only one of the great experts on the evolution of life on Earth and possibly elsewhere, he is a very kind and patient teacher. In addition to answering my early questions about the origins and evolution of life, he later read and critiqued a draft of this manuscript and pointed out to me various errors and confusions, egregious and otherwise, that had crept in. I feel particularly fortunate to have had the benefit of Andy's expertise and I only hope that I can return the favor. Finally, at another speaking engagement (this time related to my scientific research) organized by the National Academy of Sciences, I had the good fortune to meet and listen to Daniel Schrag, of Harvard, one of the originators of the Snowball Earth hypothesis. He was kind enough to provide me with several of his own articles on this subject, which nicely complemented what I could find in the popular literature.

I asked several colleagues to look at various versions of the manuscript. My friend and frequent collaborator Frank Wilczek, who, besides his great knowledge of physics, has, to first approximation, read widely on almost every other subject in science, made a variety of useful suggestions. I also thank Freeman Dyson, for pointing out several useful sources. My wonderful and wise agent, friend, and ex-editor, Susan Rabiner, was an asset in every phase of the writing of this manuscript, from the initial book proposal to the final revisions. She was unstintingly generous with her suggestions and criticisms. Finally, I would like to thank the three editors at Little, Brown, Rick Kot, Bill Phillips, and Deborah Baker, who at various times have been involved with this book. Each has made an important contribution to help guide the project along. My staff assistant, Lori Rotar, was of great help in conveying information between me and the staff at Little, Brown.

Discussions with colleagues, however, no matter how well informed, cannot provide sufficient training for a task of this sort. I was fortunate to find a wide variety of excellent monographs on most of the key subjects I felt I needed to learn about. In addition, since I particularly wanted to introduce cutting-edge results from a wide variety of areas, I benefited greatly from news reporting and review articles in various science magazines, including *Nature, Science,* and *Scientific American.*

Finally, there are two other individuals who deserve special thanks. Jan Willem Nienhuys was given my initial manuscript to help him prepare to produce the Dutch translation of this book. In a series of emails resulting in over 100 pages of comments, he peppered me with questions, checking over every calculation in the book and finding numerous typos and errors, and in the process educating me about a host of subjects. I could not have hoped to hire a researcher with the broad background and attention to detail that he provided. Lastly, I want to thank my wife, Kate. She is an incisive critic, and her encouragement upon reading my early efforts had an important impact on my writing, as do her good sense and hearty spirit in all facets of my life.

Naturally, I take responsibility for any errors that remain. I hope my many tutors will forgive any imperfect rendering of the knowledge they imparted.

❄

Since this is not, strictly speaking, a scholarly text, I did not feel it was appropriate to provide explicit references throughout. But I did want to list those books I found particularly useful, focusing for the most part, but not exclusively, on those areas outside my own fields of research in physics and cosmology, to which the interested reader might turn to learn more:

Adams, F., and G. Laughlin. *The Five Ages of the Universe*. Free Press, 1999. This recent popular book presents a thoughtful and authoritative review of the long-time evolution of stars.

Beatty, J. K., C. C. Petersen, and A. Chaikin, eds. *The New Solar System*. Cambridge University Press, 1999. This compendium of articles on all facets of the astronomy and geology of our solar system contains up-to-date information about almost every question one might have in planetary science.

Brack, André. *The Molecular Origins of Life: Assembling the Pieces of the Puzzle*. Cambridge University Press, 1998. This conference proceedings provides a useful and up-to-date introduction to a variety of areas, including interesting discussions of comets and the origins of life.

Celebrating the Neutrino. Los Alamos Science, no. 25, 1997. A nice review, in a single magazine, of many aspects of neutrino physics. I found the discussions of supernova physics here particularly useful.

Clayton, Donald D. *Principles of Stellar Evolution and Nucleosynthesis*. University of Chicago Press, 1983. I have long turned to this classic graduate-level text (originally published in 1968) as a reference to the structure and evolution of stars.

Eddington, Arthur Stanley. *The Internal Constitution of the Stars*. Cambridge University Press, 1926. The classic text that laid the basis of modern stellar astrophysics, and also presented a model of scholarly writing that few others can match.

Fortey, Richard. *Life: A Natural History of the First Four Billion Years of Life on Earth*. Knopf, 1998. A wonderful read, and an incredibly good introduction to almost every topic associated with the evolution of the Earth and the life it hosts.

Ingraham, Lloyd L. *The Biochemistry of Dioxygen.* Cambridge University Press, 1985. A comprehensive and detailed review of the basic biochemistry of oxygen.

Jakosky, Bruce. *The Search for Life on Other Planets.* Cambridge University Press, 1998. A succinct and up-to-date review of the central issues associated with the origin of life, including those factors that relate to the possible existence of life elsewhere in the solar system, and beyond.

Levi, Primo. *The Periodic Table.* Trans. Raymond Rosenthal. Schocken, 1984. The last chapter presents the recent life history of a carbon atom in the spirit of this book, but with a literary style that undoubtedly puts mine to shame.

Margulis, Lynn. *Early Life.* Jones and Bartlett, 1984. This presents a succinct review of key aspects associated with the origin of life, and nicely presents Margulis's thesis regarding the origins of organelles within eukaryotic cells.

Mason, Stephen F. *Chemical Evolution, Origin of the Elements, Molecules, and Living Systems.* Clarendon Press, 1991. A masterpiece of scholarship. I have rarely found in one place so authoritative and readable an introduction to so many diverse fields. The author presents a coherent picture of chemical evolution in many settings, offering important discussions in astronomy, geology, geochemistry, cosmochemistry, biology, and biochemistry.

Shklovskii, Iosif S. *Stars, Their Birth, Life, and Death.* Freeman, 1978. While by now out of date in several areas, this popular treatment of various aspects of stellar formation provides a good introduction to many of the central ideas in this field.

Thomas, P., C. Chyba, and C. McKay, eds. *Comets and the Origin and Evolution of Life.* Springer-Verlag, 1997. Similar in style and substance to the Brack book, it presents a complementary view of a number of issues in this area.

Wallace, Robert A., Jack L. King, and Gerald P. Sanders. *Biosphere: The Realm of Life*. Scott, Foresman, 1984. An introductory college-level text to modern evolutionary biology. It provides good general discussions of the basic energetics of cell biology, and the key metabolic processes at the basis of life.

INDEX

Acasta River, 160
accidents, 17, 19–20
acid rain, 166, 167, 169, 174, 234–35
adenosine diphosphate (ADP), 181
adenosine monophosphate (AMP), 181
adenosine triphosphate (ATP), 181, 191,
    192, 196, 199, 204–6, 218, 240
algae, 179, 210; blue-green (*see* cyanobac-
    teria)
Allen, Woody, 269
Allende meteorite, 141, 142
aluminum, 139, 142, 143, 150–51, 153
aluminum oxide, 149
Alvin, 176
Amhairghin, 236, 242
amino acids, 124, 174, 186, 187; handed-
    ness of, 125, 175–76
ammonia, 124, 174
Andromeda galaxy, 131, 132, 262–63
animals, 212, 226; origins of, 219–20
Antarctica: listening to CBR in, 67–68;
    meteorites recovered from, 124–26, 140,
    184
antihydrogen, 23, 24–25
antimatter, 23–28, 60; asymmetry of mat-
    ter and, 25, 26–27, 31–32, 35–45, 281–82;
    mass of, 27

antineutrinos, 29, 163
antiparticles, 23–28, 31–32, 35–45, 60. *See
    also specific antiparticles*
antiprotons, 23, 24, 35
antiquarks, 43–44, 46, 51
archaea, 188–89, 190–91, 205, 208
argon, 154
arrow of time, 36–40
Asimov, Isaac, 34
asteroid belt, 125, 145, 146, 151, 153, 227–28.
    See also meteorites; meteors
asteroids, 160; future collisions with, 246,
    252; K/T impact and, 231–35, 244, 246;
    Mars bombarded by, 147–48
astronomy, 112–14, 157–58
atmosphere (of Earth), 181; bombardment
    by objects of extraterrestrial origin and,
    175, 234; changes in composition of, 174,
    183, 197, 201–4, 209–12, 218, 219, 221,
    223–27, 231, 251, 253–56; formation of,
    144–47, 152–54, 161–67; in future, 257;
    life-forms' effects on, 183, 197, 201–4,
    209–12; number of oxygen atoms in, 241;
    oxygen atom released into, 169; radia-
    tion in, 195, 210; recycling time for oxy-
    gen atoms in, 239–40, 241–42. *See also*
    greenhouse gases; *specific constituents*

phosphates and, 180; photosynthesis and, 183, 185, 200, 201, 204–5, 209–10, 240, 242; as potential threat to life, 177, 200–201, 204, 207–8, 210–11, 218; reactivity of, 189–90, 200–202, 206, 210–11; respiration and, 172, 173–74, 205–10, 219, 221, 242, 249, 250, 266; in stars, 109–11, 121, 122; ultimate creation of, 134–36; vital to process of life, 249, 250
oxygen-15, 121
oxygen-16, 107, 109
ozone, 185, 189, 203, 210

palladium, 142
Pangaea, 159, 228
panspermia, 140–41, 175–76
particle accelerators, 20–23, 40, 42, 47, 50–51
particles, 14–15, 40; antiparticles and, 23–28, 31–32, 35–45; changes in identity of, 22–23, 43, 60, 61; collisions of, in early universe, 21–23; laws describing (*see* quantum mechanics); sparticles and, 47. *See also specific particles*
Pascal, Blaise, 65
past: atomic connection to, 241–42; time's arrow and, 37–38
Pauling, Linus, 207
Paulos, John Allen, 241
Periodic Table, The (Levi), 238, 240
Perlmutter, Saul, 273–74
Permian era, 228–29, 231
phosphates, 180–81
phosphorescence, 196
phosphorus, 180
photolysis, 195
photons, 22–23, 26, 28, 61, 74–75, 79, 84, 114, 116, 122; decay of protons into, 28, 30; electromagnetism and, 56; electrons knocked free by, 72–73; ratio of protons to, 26–27, 31–32, 44
photoplankton, 251
photosynthesis, 195–99, 211, 219, 220, 221, 226, 234, 239–40, 242, 249–50; in early bacteria, 183, 185, 189, 195–200, 209–10; electron transport and, 193, 195, 196; light and dark reactions in, 196; oxygen

produced in, 185, 200, 201, 204–5, 209–10; three stages of, 199
Photosystem I (PS I) reactions, 199, 204
Photosystem II (PS II) reactions, 199, 204
phylogenetic tree of life, 188
Pilbara rocks, 183
pions, 38–39
Planck, Max, 279
planetesimals, 119, 126, 144, 146, 150–53, 155
planets, 161, 261; atmosphere of, 126–29, 132; discovery of, 157–58; formation of, 119–24, 126–28, 139–52; motions of, 245. *See also specific planets*
plankton, 185, 210, 251
plants, 210, 212, 226, 234
plate tectonics, 158–59, 163, 167–70, 176, 198, 203, 220, 221, 253, 255
Politzer, David, 40–41
porphyrin rings, 193, 195–96, 207
positrons, 24, 45, 46, 65–66, 282–83; decay of protons into, 28, 30
potassium, 154, 163
Precambrian era, 179, 219, 220
predictability, 245–46
pressure, 82; struggle between gravity and, 16, 71, 77, 80, 84–89, 93–94, 96–98, 101, 104–6, 108, 109; thermal energy and, 84, 85, 88–89
primordial density enhancements (gas clouds), 68–69, 71, 75–76, 79–89, 130–31; COBE detector and, 68–69; collisions of atoms in, 82–83, 89; dissipated by radiation pressure, 71; dissipation of energy of motion in, 82–84; as gas, 82; gas making up interstellar space today vs., 83; gravitational collapse of, 75–76, 80–89, 96–98; ionized atoms in, 84, 97, 98; lack of uniform density in, 81; luminosity of, 88, 97–98; magnetic field in, 84, 86; rotational velocity of, 86; thermal energy in, 84, 85, 87–89, 96–97
Principle of Copernican Time, 247–48
prokaryotes, 186–88, 209, 211, 212
proteins, 187, 207
proton-decay detectors, 45–49

Virus X
Tracking the New Killer Plagues
by Frank Ryan, M.D.

"A very readable and disturbing book. . . . Dr. Ryan writes well in a difficult technical field, weaving the technicalities of scientific history, medicine, molecular biology, and evolution into the human narratives of a sequence of epidemics, with victims, heroes (the medical teams), and villains (politicians, more or less)."
— John R. G. Turner, *New York Times Book Review*

"At times thrilling and surprising. . . . Not many books — fiction or nonfiction — have this cosmic ability."
— Miroslav Holub, *Los Angeles Times*

The River
A Journey to the Source of HIV and AIDS
by Edward Hooper

"Is AIDS a disaster inadvertently brought on by humans in the early testing of a polio vaccine? . . . This remarkable book offers tantalizing clues to revive and expand this theory."
— Lawrence K. Altman, M.D., *New York Times*

"A work of spine-chilling forensic imagination. . . . It's brilliantly researched, passionately written, and an astonishingly easy and thoroughly gripping read."   — Geoffrey Robertson, *Observer* Books of the Year